Jacob Fuchs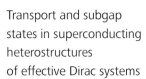

Transport and subgap
states in superconducting
heterostructures
of effective Dirac systems

T0134341

# Transport and subgap states in superconducting heterostructures of effective Dirac systems

Dissertation zur Erlangung des Doktorgrades der Naturwissenschaften (Dr. rer. nat.)
der Fakultät für Physik der Universität Regensburg
vorgelegt von

Jacob Fuchs

aus Bietigheim-Bissingen
im Juni 2022

Die Arbeit wurde von Prof. Dr. Klaus Richter angeleitet.
Das Promotionsgesuch wurde am 11. Mai 2022 eingereicht.
Das Kolloquium fand am 27. Juli 2022 statt.

Prüfungsausschuss:    Vorsitzender:       Prof. Dr. Dieter Weiss
                      1. Gutachter:       Prof. Dr. Klaus Richter
                      2. Gutachter:       Prof. Dr. Milena Grifoni
                      weiterer Prüfer:    Prof. Dr. Christoph Lehner

Dissertationsreihe der Fakultät für Physik der Universität Regensburg,
Band 57

Herausgegeben vom Präsidium des Alumnivereins der Physikalischen Fakultät:
Klaus Richter, Andreas Schäfer, Werner Wegscheider

Jacob Fuchs

# Transport and subgap
# states in superconducting
# heterostructures of
# effective Dirac systems

Universitätsverlag Regensbu

Bibliografische Informationen der Deutschen Bibliothek.
Die Deutsche Bibliothek verzeichnet diese Publikation
in der Deutschen Nationalbibliografie. Detailierte bibliografische Daten
sind im Internet über http://dnb.ddb.de abrufbar.

1. Auflage 2022
© 2022 Universitätsverlag, Regensburg
Leibnizstraße 13, 93055 Regensburg

Konzeption: Thomas Geiger
Umschlagentwurf: Franz Stadler, Designcooperative Nittenau eG
Layout: Jacob Fuchs
Druck: Docupoint, Magdeburg
ISBN: 978-3-86845-171-9

Weitere Informationen zum Verlagsprogramm erhalten Sie unter:
www.univerlag-regensburg.de

# Transport and subgap states in superconducting heterostructures of effective Dirac systems

**Dissertation**
zur Erlangung des Doktorgrades
der Naturwissenschaften (Dr. rer. nat.)
der Fakultät für Physik
der Universität Regensburg

vorgelegt von
**Jacob Fuchs**
aus Bietigheim-Bissingen

im Jahr 2022

Promotionsgesuch eingereicht am:   11.05.2022
Die Arbeit wurde angeleitet von:      Prof. Dr. Klaus Richter

Dedicated to all friends and family members
who took care of me in the hospital
and during recovering from my disease.

# Contents

# Introduction

Throughtout this thesis, we will investigate two different materials, graphene and topological insulators (TIs). At first sight, these two are very different: Graphene is a 2D material which consists of a single layer of carbon atoms [1, 2]. TIs on the other hand, are either 2D quantum wells or 3D materials made from metal compounds [3], HgCdTe, HgTe, $Bi_{1-x}Sb_x$, $Bi_2Se_3$, or $Bi_2Te_3$ for example. However, both of them provide realizations of massless Dirac electrons in solid state systems: In graphene, electrons near the $K$ and $K'$ points behave like massless Dirac electrons, whereas for TIs, the surface states can be described with the 2D massless Dirac equation while the bulk is insulating. Thus, we call both of them *effective Dirac systems*—"effective", because these descriptions are only valid in the low energy limit or for the surface states, respectively.

Here, we will consider *superconducting heterostructures,* i.e., hybrid structures of these effective Dirac systems with superconductors. Superconductors are materials with zero electrical resistence [4]. An intriguing phenomenon in such systems is the proximity effect [5]—in contact with a superconductor, usually nonsuperconducting materials can become superconducting themselves. In the first Chapter, we use this effect implicitly to introduce superconductivity in bilayer graphene (BLG), i.e., two layers of graphene stacked on top of each other. In the other Chapters, superconductivity is induced only in a part of the system leading to heterostructures containing normal and superconducting parts. Such systems show a lot of interesting phenomena, like Andreev reflection (AR) [6], Josephson currents and Andreev bound states [5, 7–11], or Yu-Shiba-Rusinov (YSR) states [12–16], as well as topological superconductivity and Majorana zero modes [17–21], which were discovered more recently and predicted to enable fault-tolerant quantum computation [19]. This explains why these heterostructures are intensively investigated theoretically and experimentally.

In the first Chapter, we investigate superconducting bilayer graphene (SBLG), where an adatom is chemisorbed onto a carbon atom. For the adatom, we use hydrogen since this is the most common impurity, although other adatoms like fluorine should give similar results. Superconducting systems are characterized by a superconducting gap, which appears to be, at the same time, the order parameter of the phase transition [4]. However, the adatoms form impurities which can lead to the formation of *subgap states,* i.e., bound states with energies inside the superconducting gap. These states are called Yu-Shiba-Rusinov (YSR) states [12–16]. Our goal is to confirm the existence of YSR states in this system and to calculate their spectrum. On the one hand, this could motivate an experimental investigation since YSR states have experimentally been confirmed to exist in graphene grain boundaries [22]. On the other hand, it has been shown that the existence of YSR states in superconducting graphene heterostructures influences spin relaxation [23, 24].

In the next Chapters, we switch to heterostructures with nanowires made from 3D TIs and investigate their *transport* behavior.

Heterostructures involving both, normal and superconducting parts, exhibit a prominent *transport process* called "Andreev reflection" (AR). Here, an electron coming in through the normal part is reflected as a hole, and at the same time, a Cooper pair is formed in the superconductor. The presence of a second normal contact enables another process called "crossed Andreev reflection" (CAR), where the outgoing hole is located in the other normal contact. This is of particular interest since its reversed process splits a Cooper pair into two entangled electrons in different contacts [25–27]. In the second Chapter, we propose a T junction device made out of 3D TI nanowires and show that this system allows for the observation of CAR by numerically simulating its transport properties.

In the third Chapter, we look at Josephson junctions made from 3D TI nanowires. Since the superconducting TI nanowires are predicted to be topolgical and host Majorana zero modes [28, 29], these systems should give rise to the fractional Josephson effect amongst other things [for a recent experiment on this subject, see 30]. Josephson junctions are superconducting heterostructures where a weak link, for example a normal metal or an insulator, is sandwiched in between two superconductors. Josephson [7] predicted that a supercurrent between the two superconducting contacts is present despite the nonsuperconducting regions in between, earning him the Nobel prize in 1973 [31]. While this current is a *transport* process, it can be explained by *subgap states* called Andreev bound states (ABSs) [5, 8–11]; this constitutes a link between the two phenomena, transport and subgap states, investigated in this thesis. We aim to analytically calculate the critical current, which is the maximum Josephson current, and investigate its behavior in dependence of a magnetic field parallel to the nanowire. Recent experiments revealed oscillations of the critical current with an unexpected periodicity. With our theoretical model, we propose an explanation for this behavior and, through a semiclassical analysis, reveal the physical origin of these oscillations.

# Chapter 1.

# Yu-Shiba-Rusinov states in superconducting bilayer graphene

The results in this Chapter have been published by Barth, Fuchs, and Kochan [24]. The analytical calculation of the Yu-Shiba-Rusinov (YSR) states has been performed by Jacob Fuchs with the assistance of Denis Kochan.

## 1.1. Introduction

### Yu-Shiba-Rusinov states

Between 1965 and 1968, Yu [12], Shiba [13], and Rusinov [14, 15] showed independently that a (single) magnetic impurity in an $s$-wave superconductor can host bound states. In honor of their discoverers, these states are nowadays called Yu-Shiba-Rusinov (YSR) states or just simply Shiba states. For example, Shiba [13] investigated an $s$-wave BCS superconductor with a classical magnetic impurity coupled to the conduction electrons. The Hamiltonians of the superconductor and the impurity read

$$H_{BCS} = \sum_{k,\alpha} \varepsilon_k c_{k,\alpha}^{\dagger} c_{k,\alpha} + \Delta_0 \sum_k (c_{k,+}^{\dagger} c_{-k,-}^{\dagger} + c_{-k,-} c_{k,+}) \tag{1.1}$$

and

$$H_{ex} = \frac{1}{2N} \sum_{k,k',\alpha,\beta} J c_{k,\alpha}^{\dagger} (\sigma_{\alpha,\beta} \cdot S) c_{k',\beta}, \tag{1.2}$$

respectively; here, $J$ describes the interaction strength, $S = |S|$ is the impurity spin and $N$ the system size. Shiba considered a classical spin, neglecting all quantum mechanical effects, by investigating the limit $J \rightarrow 0$, $S \rightarrow \infty$ with the constraint

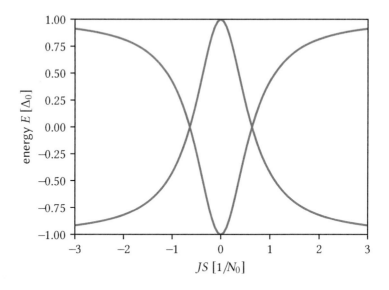

*Figure 1.1. Example of Yu-Shiba-Rusinov states.* YSR states of a classical spin with a spherically symmetric exchange as calculated by Shiba [13], see Eq. (1.3).

$JS = const.$ Through the Green's function and $T$ matrix [13, 32],[1,2] he found subgap states at energies

$$E = \pm\Delta_0 \frac{1 - (JS\pi N_0/2)^2}{1 + (JS\pi N_0/2)^2},$$  (1.3)

where $N_0$ is the normal-state density of states (DOS) at the Fermi energy. Figure 1.1 shows how the energy of these states depends on the product $JS$ given a constant normal-state DOS $N_0$. Note that this energy spectrum shows a zero crossing ($E = 0$) for some critical value $J_c$ of $J$,[3] where the nature of the ground states typically switches between a singlet and doublet state [16].

It took several years until YSR states could be realized experimentally. First experiments focused on magnetic impurities in superconductors [33–35]. Later, YSR states have been reported in magnetic moments [36] and atomic chains adsorbed onto a superconductor [37–41], magnetic islands [42, 43], and molecular junctions [44]. In

---

[1]A similar approach was used by Rusinov [15]. Yu [12] and Rusinov [14] derived similar results by Bogoliubov transformation [see also 32].

[2]In this work, we use a similar method; compare Section 1.4.2.

[3]We use $J$ here, and not $JS$, since the spin $S$ is constant in real physical systems.

the last years, YSR states have gained particular interest in the search for topological phase transitions [16, 37] and Majorana modes and their applications in quantum computing [45–48].

## Graphene

Graphene [1, 2, 49–53] is a single layer of graphite, i.e., a one atom thick sheet of carbon atoms arranged in a 2D honeycomb lattice. While the band structure is known since the 1940s [54], the first experimental realization was reported in 2004 by Novoselov et al. [55]: Following an idea by Andre Geim, they peeled off a layer from graphite using an adhesive tape and transferred it to a silicon substrate [55, 56]. This marked the starting point of an thriving research on graphene and other 2D materials [57]. In 2010, Novoselov and Geim have been awarded the Nobel Prize for their achievement [58–60].

Of special interest are heterostructures of graphene and bilayer graphene (BLG) [61–65] with other 2D materials [66–69]. These systems offer, for example, high mobility [70] or large spin diffusion lengths [71] which makes them suitable for spintronics applications. In these heterostructures, the proximity effect can introduce additional effects in graphene like spin-orbit coupling or magnetic exchange interaction [70]. Proximity induced superconductivity is also of special interest [72–75] and has already been experimentally verified [76–84].

## Yu-Shiba-Rusinov states in graphene

YSR states in graphene have first been computed by Lado and Fernández-Rossier [85]. They investigated graphene on top of a superconductor functionalized by a hydrogen atom, where electron-electron interactions induce local magnetic moments. They showed that hydrogenated graphene exhibits subgap states by calculating the corresponding YSR state spectrum. Kochan et al. [23] studied YSR states using a method similar to the one employed in this Chapter and also investigated their effect on the quasiparticle spin relaxation.

YSR states have also been experimentally observed in graphene grain boundaries by Cortés-del Río et al. [22] using scanning tunneling microscopy and spectroscopy. Graphene grain boundaries are boundaries between graphene domains with a different orientation of the lattice, where under-coordinated carbon atoms exist. Cortés-del Río et al. deposited lead (Pb) islands on a graphene surface to induce superconductivity in the graphene layer. Using scanning tunneling spectroscopy, they measured the local density of states (LDOS) which reveals the induced superconducting gap as well as the subgap YSR states near the graphene grain boundaries. The experimental observation of YSR states in this system also provides a proof of the existence of magnetic moments in such boundaries since magnetic moments are necessary for the formation of YSR states [22]. Note that the observation of YSR states in chemically pure systems is also exceptional.

**Outline**

In this work, we want to investigate superconducting Bernal-stacked bilayer graphene (SBLG) functionalized by a single hydrogen atom and explore YSR states in such systems. We use realistic parameters from density functional theory calculations such that our predictions can be put to the test in experiments.

First, we introduce the theoretical model for bilayer graphene (BLG) and deduce the Bogoliubov-de Gennes Hamiltonian of SBLG in Section 1.2. We proceed by calculating the Green's function of SBLG in Section 1.3. With that, we calculate the YSR states in Section 1.4. The results are shown and discussed in Section 1.4.5. Appendices A and B contain some mathematical details.

## 1.2. Model of superconducting bilayer graphene

First, we need to define the Hamiltonian for SBLG and derive the corresponding Bogoliubov-de Gennes Hamiltonian. For this, we follow the procedure of Kochan et al. [23] for single layer graphene and extend it to BLG.

### 1.2.1. Lattice structure

The top view of the lattice structure of Bernal-stacked (or *AB*-stacked) bilayer graphene is shown in Fig. 1.2. Two graphene sheets, denoted by 1 and 2, individually consist of two triangular sublattices, denoted by *A* and *B*. The primitive vectors of these sublattices are given by

$$\boldsymbol{a}_1 = (a, 0) \quad \text{and} \quad \boldsymbol{a}_2 = \left(\tfrac{1}{2}a, \tfrac{1}{2}\sqrt{3}a\right) \tag{1.4}$$

with $a = 0.246\,\text{nm}$ being the lattice constant. The directed distances from an *A* atom to its three neighbouring *B* atoms are given by

$$\boldsymbol{\delta}_1 = \left(0, \tfrac{1}{\sqrt{3}}a\right), \quad \boldsymbol{\delta}_2 = \left(-\tfrac{1}{2}a, -\tfrac{1}{2\sqrt{3}}a\right), \quad \text{and} \quad \boldsymbol{\delta}_3 = \left(\tfrac{1}{2}a, -\tfrac{1}{2\sqrt{3}}a\right) \tag{1.5}$$

such that the distance between two neighbouring carbon atoms is $a_{\text{cc}} = a/\sqrt{3} = 0.142\,\text{nm}$. While the second graphene layer has the same lattice structure as the first

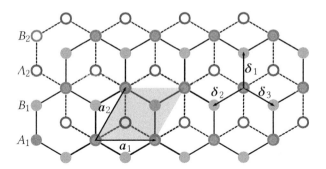

*Figure 1.2. BLG lattice. A and B depict the two sublattices for the lower (1) and upper (2) graphene layers. The brown rhombus shows the unit cell, $\boldsymbol{a}_{1,2}$ the primitive vectors, and $\boldsymbol{\delta}_{1,2,3}$ the vectors from an A atom to its nearest neighbors.*

one, it is shifted by $-\boldsymbol{\delta}_1$ such that any $B_2$ atom is above an $A_1$ atom (compare Fig. 1.2). Therefore, the positions of all atoms are

$$\mathbf{P}_{A_1} = \mathbf{P} = \{na_1 + ma_2\}, \qquad \mathbf{P}_{B_1} = \boldsymbol{\delta}_1 + \mathbf{P} = \{\boldsymbol{\delta}_1 + na_1 + ma_2\}, \qquad (1.6)$$

$$\mathbf{P}_{A_2} = -\boldsymbol{\delta}_1 + \mathbf{P}, \qquad \mathbf{P}_{B_2} = \mathbf{P} = \mathbf{P}_{A_1}. \qquad (1.7)$$

Note that the reciprocal lattice is also a honeycomb lattice with the primitive vectors

$$\boldsymbol{b}_1 = \left( \tfrac{2\pi}{a}, -\tfrac{1}{\sqrt{3}} \tfrac{2\pi}{a} \right) \quad \text{and} \quad \boldsymbol{b}_2 = \left( 0, \tfrac{2}{\sqrt{3}} \tfrac{2\pi}{a} \right). \qquad (1.8)$$

The coordinates of the $\boldsymbol{K}$ and $\boldsymbol{K}'$ points are given by

$$\boldsymbol{K} = \left( \tfrac{4\pi}{3a}, 0 \right) \quad \text{and} \quad \boldsymbol{K}' = -\boldsymbol{K}. \qquad (1.9)$$

## 1.2.2. Tight binding Hamiltonian of bilayer graphene

Let $X_j(\boldsymbol{R}, \sigma)$ be the annihilation operator corresponding to a particle with spin $\sigma$ on the sublattice $X = A, B$ of the graphene layer $j = 1, 2$ at the site with the position $\boldsymbol{R} \in \mathbf{P}_{X_j}$. The number of lattice cells is denoted by $N$. To write the tight binding Hamiltonian, we use the convention that the summation

$$\sum_R \equiv \sum_{R \in P} \qquad (1.10)$$

runs over all lattice cells $\mathbf{P}$ (except when explicitly noted otherwise) and the summation

$$\sum_{\langle R,R'+\delta_1\rangle} \tag{1.11}$$

runs over all cells $R, R' \in \mathbf{P}$ such that $R$ and $R' + \delta_1$ are nearest neighbours.

The tight binding Hamiltonian can be divided into four parts,

$$H = H_0 + H_1 + H_\mu + H_\Delta, \tag{1.12}$$

where

$$\begin{aligned}
H_0 = & -\gamma_0 \sum_{\langle R,R'+\delta_1\rangle,\sigma} \left(A_1^\dagger(R,\sigma)B_1(R'+\delta_1,\sigma) + B_1^\dagger(R'+\delta_1,\sigma)A_1(R,\sigma)\right) \\
& -\gamma_0 \sum_{\langle R-\delta_1,R'\rangle,\sigma} \left(A_2^\dagger(R-\delta_1,\sigma)B_2(R',\sigma) + B_2^\dagger(R',\sigma)A_2(R-\delta_1,\sigma)\right)
\end{aligned} \tag{1.13}$$

describes the two graphene layers within the nearest-neighbour approximation,

$$H_1 = \gamma_1 \sum_{R,\sigma} \left(A_1^\dagger(R,\sigma)B_2(R,\sigma) + B_2^\dagger(R,\sigma)A_1(R,\sigma)\right) \tag{1.14}$$

captures the interlayer coupling between them,

$$\begin{aligned}
H_\mu = & -\mu \sum_{R,\sigma} \left(A_1^\dagger(R,\sigma)A_1(R,\sigma) + B_1^\dagger(R+\delta_1,\sigma)B_1(R+\delta_1,\sigma)\right. \\
& \left. + A_2^\dagger(R-\delta_1,\sigma)A_2(R-\delta_1,\sigma) + B_2^\dagger(R,\sigma)B_2(R,\sigma)\right)
\end{aligned} \tag{1.15}$$

stands for the chemical potential (energy is measured from the Fermi level), and

$$\begin{aligned}
H_\Delta = & \Delta \sum_R \left(A_1^\dagger(R,-)A_1^\dagger(R,+) + B_1^\dagger(R+\delta_1,-)B_1^\dagger(R+\delta_1,+)\right. \\
& \left. + A_2^\dagger(R-\delta_1,-)A_2^\dagger(R-\delta_1,+) + B_2^\dagger(R,-)B_2^\dagger(R,+)\right) \\
& + \Delta^* \sum_R \left(A_1(R,+)A_1(R,-) + B_1(R+\delta_1,+)B_1(R+\delta_1,-)\right. \\
& \left. + A_2(R-\delta_1,+)A_2(R-\delta_1,-) + B_2(R,+)B_2(R,-)\right)
\end{aligned} \tag{1.16}$$

governs the superconducting $s$-wave coupling. This is the McClure-Slonczewski-Weiss parametrization [24, 86–89], where the "skew" couplings, conveniently labeled as $\gamma_{3,4}$, have been omitted to simplify the model. The atoms $A_1$ and $B_2$, coupled via $\gamma_1$, are called dimer sites, whereas the atoms $B_1$ and $A_2$, not coupled via $\gamma_1$, are called nondimer sites. In this work, the graphene nearest neighbour hopping is taken to

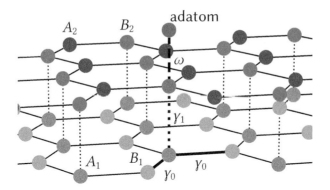

*Figure 1.3. Tight-binding model of BLG with an adatom chemisorbed onto a dimer site. The BLG lattice is the same as in Fig. 1.2. Some of the hoppings are drawn in bold and annotated with the corresponding hopping parameters.*

be $\gamma_0 = 2.6\,\text{eV}$, the interlayer hopping $\gamma_1 = 0.339\,\text{eV}$, and the distance between the graphene layers $c = 0.335\,\text{nm}$ [24]. The tight binding model and its hoppings are depicted in Fig. 1.3 (note that we already included the adatom from Section 1.4.1 there).

### 1.2.3. Tight binding Hamiltonian in Bloch basis

Next, one has to express the Hamiltonian within the Bloch basis $|X_j, \mathbf{k}, \sigma\rangle$ instead of the real-space tight binding basis $|X_j, \mathbf{R}, \sigma\rangle$. The corresponding transformation of the basis states is given by

$$|X_j, \mathbf{k}, \sigma\rangle = \frac{1}{\sqrt{N}} \sum_{R \in P_{X_j}} \exp(\mathrm{i}\mathbf{k} \cdot \mathbf{R}) |X_j, \mathbf{R}, \sigma\rangle, \tag{1.17}$$

whereas the transformation of the annihilation operators reads

$$X_j(\mathbf{k}, \sigma) = \frac{1}{\sqrt{N}} \sum_{R \in P_{X_j}} \exp(\mathrm{i}\mathbf{k} \cdot \mathbf{R}) X_j(\mathbf{R}, \sigma). \tag{1.18}$$

For any $\mathbf{k}$ in the first Brillouine zone, one has

$$\sum_{R \in P_{X_j}} \exp(\mathrm{i}\mathbf{k} \cdot \mathbf{R}) = N\delta_{k,0} \quad \text{for all } X \text{ and } j \tag{1.19}$$

and

$$\sum_{\langle R,R'+\delta_1 \rangle} \exp(\mathrm{i}\boldsymbol{k} \cdot \boldsymbol{R} - \mathrm{i}\boldsymbol{k}' \cdot (\boldsymbol{R}' + \boldsymbol{\delta}_1))$$

$$= \sum_{\langle R-\delta_1, R' \rangle} \exp(\mathrm{i}\boldsymbol{k} \cdot (\boldsymbol{R} - \boldsymbol{\delta}_1) - \mathrm{i}\boldsymbol{k}' \cdot \boldsymbol{R}') = N\delta_{\boldsymbol{k},\boldsymbol{k}'} f(\boldsymbol{k}), \qquad (1.20)$$

where

$$f(\boldsymbol{k}) = \exp(-\mathrm{i}\boldsymbol{k} \cdot \boldsymbol{\delta}_1) + \exp(-\mathrm{i}\boldsymbol{k} \cdot \boldsymbol{\delta}_2) + \exp(-\mathrm{i}\boldsymbol{k} \cdot \boldsymbol{\delta}_3) \qquad (1.21)$$

is the nearest-neighbor structural function of the graphene lattice. Thus, the Hamiltonian in the Bloch basis is given by

$$H_0 = -\gamma_0 \sum_{\boldsymbol{k},\sigma} f(\boldsymbol{k}) \big( A_1^\dagger(\boldsymbol{k},\sigma)B_1(\boldsymbol{k},\sigma) + A_2^\dagger(\boldsymbol{k},\sigma)B_2(\boldsymbol{k},\sigma) \big)$$

$$- \gamma_0 \sum_{\boldsymbol{k},\sigma} f^*(\boldsymbol{k}) \big( B_1^\dagger(\boldsymbol{k},\sigma)A_1(\boldsymbol{k},\sigma) + B_2^\dagger(\boldsymbol{k},\sigma)A_2(\boldsymbol{k},\sigma) \big), \qquad (1.22)$$

$$H_1 = \gamma_1 \sum_{\boldsymbol{k},\sigma} \big( A_1^\dagger(\boldsymbol{k},\sigma)B_2(\boldsymbol{k},\sigma) + B_2^\dagger(\boldsymbol{k},\sigma)A_1(\boldsymbol{k},\sigma) \big), \qquad (1.23)$$

$$H_\mu = -\mu \sum_{\boldsymbol{k},\sigma} \big( A_1^\dagger(\boldsymbol{k},\sigma)A_1(\boldsymbol{k},\sigma) + B_1^\dagger(\boldsymbol{k},\sigma)B_1(\boldsymbol{k},\sigma) \big)$$

$$- \mu \sum_{\boldsymbol{k},\sigma} \big( A_2^\dagger(\boldsymbol{k},\sigma)A_2(\boldsymbol{k},\sigma) + B_2^\dagger(\boldsymbol{k},\sigma)B_2(\boldsymbol{k},\sigma) \big), \qquad (1.24)$$

and

$$H_\Delta = \Delta \sum_{\boldsymbol{k}} \big( A_1^\dagger(-\boldsymbol{k},-)A_1^\dagger(\boldsymbol{k},+) + B_1^\dagger(-\boldsymbol{k},-)B_1^\dagger(\boldsymbol{k},+) \big)$$

$$+ \Delta \sum_{\boldsymbol{k}} \big( A_2^\dagger(-\boldsymbol{k},-)A_2^\dagger(\boldsymbol{k},+) + B_2^\dagger(-\boldsymbol{k},-)B_2^\dagger(\boldsymbol{k},+) \big)$$

$$+ \Delta^* \sum_{\boldsymbol{k}} \big( A_1(\boldsymbol{k},+)A_1(-\boldsymbol{k},-) + B_1(\boldsymbol{k},+)B_1(-\boldsymbol{k},-) \big)$$

$$+ \Delta^* \sum_{\boldsymbol{k}} \big( A_2(\boldsymbol{k},+)A_2(-\boldsymbol{k},-) + B_2(\boldsymbol{k},+)B_2(-\boldsymbol{k},-) \big). \qquad (1.25)$$

### 1.2.4. Bogoliubov-de Gennes Hamiltonian

Introducing the Nambu-like field operators

$$
\begin{aligned}
C(\boldsymbol{k}) = \big( & A_1(\boldsymbol{k},+), B_1(\boldsymbol{k},+), A_2(\boldsymbol{k},+), B_2(\boldsymbol{k},+), \\
& A_1^\dagger(-\boldsymbol{k},-), B_1^\dagger(-\boldsymbol{k},-), A_2^\dagger(-\boldsymbol{k},-), B_2^\dagger(-\boldsymbol{k},-) \big)^T
\end{aligned}
\tag{1.26}
$$

$$
\begin{aligned}
C^\dagger(\boldsymbol{k}) = \big( & A_1^\dagger(\boldsymbol{k},+), B_1^\dagger(\boldsymbol{k},+), A_2^\dagger(\boldsymbol{k},+), B_2^\dagger(\boldsymbol{k},+), \\
& A_1(-\boldsymbol{k},-), B_1(-\boldsymbol{k},-), A_2(-\boldsymbol{k},-), B_2(-\boldsymbol{k},-) \big)
\end{aligned}
\tag{1.27}
$$

and utilizing the fermionic commutation relations

$$
\{X_j^\dagger(\boldsymbol{k},\sigma), X_{j'}'(\boldsymbol{k}',\sigma')\} = \delta_{X,X'}\delta_{j,j'}\delta_{\boldsymbol{k},\boldsymbol{k}'}\delta_{\sigma,\sigma'}
\tag{1.28}
$$

as well as the relation $f(-\boldsymbol{k}) = f^*(\boldsymbol{k})$, the Hamiltonian given by Eq. (1.12) can be rearranged to

$$
H = \sum_k C^\dagger(\boldsymbol{k})\mathcal{H}(\boldsymbol{k})C(\boldsymbol{k}) - \sum_k 4\mu.
\tag{1.29}
$$

In the following, the constant $\sum_k 4\mu$, solely contributing to the ground state energy, will be dropped. Here,

$$
\mathcal{H}(\boldsymbol{k}) = \begin{pmatrix} h(\boldsymbol{k}) & -\Delta \\ -\Delta^* & -h(\boldsymbol{k}) \end{pmatrix}
\tag{1.30}
$$

is the Bogoliubov-de Gennes Hamiltonian of SBLG and

$$
h(\boldsymbol{k}) = \begin{pmatrix}
-\mu & -\gamma_0 f(\boldsymbol{k}) & 0 & \gamma_1 \\
-\gamma_0 f^*(\boldsymbol{k}) & -\mu & 0 & 0 \\
0 & 0 & -\mu & -\gamma_0 f(\boldsymbol{k}) \\
\gamma_1 & 0 & -\gamma_0 f^*(\boldsymbol{k}) & -\mu
\end{pmatrix}
\tag{1.31}
$$

the Hamiltonian of normal-phase BLG. Note that the upper left and lower right block in the last equation,

$$
h'(\boldsymbol{k}) = \begin{pmatrix} -\mu & -\gamma_0 f(\boldsymbol{k}) \\ -\gamma_0 f^*(\boldsymbol{k}) & -\mu \end{pmatrix},
\tag{1.32}
$$

represent the Hamiltonian of single layer graphene.

## 1.3. Green's function of superconducting bilayer graphene

### 1.3.1. Resolvent Green's function

The resolvent Green's function is defined as

$$G_0(\epsilon, \boldsymbol{k}) = [\epsilon - \mathcal{H}(\boldsymbol{k})]^{-1} = \begin{pmatrix} \epsilon - h & \Delta \\ \Delta^* & \epsilon + h \end{pmatrix}^{-1}. \tag{1.33}$$

In what follows, we use $\xi = \xi(\epsilon) = \sqrt{\epsilon^2 - |\Delta|^2}$, where the branch cut is chosen in a way such that $\xi = i\sqrt{|\Delta|^2 - \epsilon^2}$ whenever $|\epsilon| < |\Delta|$. Since $\Delta$ is proportional to the identity matrix, we have

$$\begin{aligned} \begin{pmatrix} \epsilon - h & \Delta \\ \Delta^* & \epsilon + h \end{pmatrix} \begin{pmatrix} \epsilon + h & -\Delta \\ -\Delta^* & \epsilon - h \end{pmatrix} &= \begin{pmatrix} \epsilon + h & -\Delta \\ -\Delta^* & \epsilon - h \end{pmatrix} \begin{pmatrix} \epsilon - h & \Delta \\ \Delta^* & \epsilon + h \end{pmatrix} \\ &= \begin{pmatrix} \epsilon^2 - h^2 - |\Delta|^2 & 0 \\ 0 & \epsilon^2 - h^2 - |\Delta|^2 \end{pmatrix} \\ &= \begin{pmatrix} \xi^2 - h^2 & 0 \\ 0 & \xi^2 - h^2 \end{pmatrix} \end{aligned} \tag{1.34}$$

and therefore

$$\begin{aligned} G_0(\epsilon, \boldsymbol{k}) &= \begin{pmatrix} (\epsilon + h)(\xi^2 - h^2)^{-1} & -\Delta(\xi^2 - h^2)^{-1} \\ -\Delta^*(\xi^2 - h^2)^{-1} & (\epsilon - h)(\xi^2 - h^2)^{-1} \end{pmatrix} \\ &= \begin{pmatrix} (\xi^2 - h^2)^{-1}(\epsilon + h) & -\Delta(\xi^2 - h^2)^{-1} \\ -\Delta^*(\xi^2 - h^2)^{-1} & (\xi^2 - h^2)^{-1}(\epsilon - h) \end{pmatrix}, \end{aligned} \tag{1.35}$$

where $(\xi^2 - h^2)^{-1} = (\xi \pm h)^{-1}(\xi \mp h)^{-1}$. Note that

$$g_0(\xi, \boldsymbol{k}) = [\xi - h(\boldsymbol{k})]^{-1} \tag{1.36}$$

is the resolvent Green's function of the Hamiltonian $h(\boldsymbol{k})$ of bilayer graphene.

One can also express the Green's function $G_0(\epsilon, \boldsymbol{k})$ totally in terms of $g_0(\xi(\epsilon), \boldsymbol{k})$. Namely,

$$G_0(\epsilon, \boldsymbol{k}) = \begin{pmatrix} A & B \\ C & D \end{pmatrix}, \tag{1.37}$$

where

$$A = \frac{1}{2} g_+(\xi) + \frac{\epsilon}{\xi} g_-(\xi), \qquad\qquad B = -\frac{\Delta}{2\xi} g_-(\xi), \qquad (1.38)$$

$$C = -\frac{\Delta^*}{2\xi} g_-(\xi), \qquad\qquad D = -\frac{1}{2} g_+(\xi) + \frac{\epsilon}{\xi} g_-(\xi), \qquad (1.39)$$

and

$$g_\pm(\xi) = g_0(\xi) \pm g_0(-\xi). \qquad (1.40)$$

## 1.3.2. Resolvent Green's function of bilayer graphene

Next, the resolvent Green's function $g_0(\xi, \boldsymbol{k}) = [\xi - h(\boldsymbol{k})]^{-1}$ of BLG is needed. We derive it in the same way like Kochan et al. [23]. First, note that $\xi - h(\boldsymbol{k})$ is a block matrix,

$$\xi - h(\boldsymbol{k}) = \begin{pmatrix} \xi - g(\boldsymbol{k}) & -\gamma_1 u \\ -\gamma_1 l & \xi - g(\boldsymbol{k}) \end{pmatrix} = \begin{pmatrix} A_1 & B_1 \\ C_1 & D_1 \end{pmatrix} \qquad (1.41)$$

with

$$u = \begin{pmatrix} 0 & 1 \\ 0 & 0 \end{pmatrix} \quad \text{and} \quad l = \begin{pmatrix} 0 & 0 \\ 1 & 0 \end{pmatrix} = u^T. \qquad (1.42)$$

Therefore, the following identity can be applied:

$$\begin{pmatrix} A_1 & B_1 \\ C_1 & D_1 \end{pmatrix}^{-1} = \begin{pmatrix} A_1^{-1} + A_1^{-1} B_1 S_1^{-1} C_1 A_1^{-1} & -A_1^{-1} B_1 S_1^{-1} \\ -S_1^{-1} C_1 A_1^{-1} & S_1^{-1} \end{pmatrix} = \begin{pmatrix} A_1' & B_1' \\ C_1' & D_1' \end{pmatrix} \qquad (1.43)$$

whenever $A_1$ and its Schur complement $S_1 - D_1 - C_1 A_1^{-1} B_1$ are invertible. With the abbreviations

$$\alpha = (\xi + \mu)^2 - \gamma_0^2 |f|^2 \quad \text{and} \quad \beta = \alpha^2 - (\xi + \mu)^2 \gamma_1^2, \qquad (1.44)$$

one can write

$$A_1^{-1} = (\xi - g)^{-1} = \alpha^{-1} \begin{pmatrix} \xi + \mu & -\gamma_0 f \\ -\gamma_0 f^* & \xi + \mu \end{pmatrix}, \qquad (1.45)$$

$$S_1 = \alpha^{-1} \begin{pmatrix} (\xi + \mu)\alpha & \gamma_0 f \alpha \\ \gamma_0 f^* \alpha & (\xi + \mu)(\alpha - \gamma_1^2) \end{pmatrix}, \qquad (1.46)$$

and

$$S_1^{-1} = \beta^{-1} \begin{pmatrix} (\xi + \mu)(\alpha - \gamma_1^2) & -\gamma_0 f \alpha \\ -\gamma_0 f^* \alpha & (\xi + \mu)\alpha \end{pmatrix}. \tag{1.47}$$

This leads to the following result:

$$g_0(\xi, \mathbf{k}) = \begin{pmatrix} A_1' & B_1' \\ C_1' & D_1' \end{pmatrix} \tag{1.48}$$

with

$$A_1' = \beta^{-1} \begin{pmatrix} (\xi + \mu)\alpha & -\gamma_0 f \alpha \\ -\gamma_0 f^* \alpha & (\xi + \mu)(\alpha - \gamma_1^2) \end{pmatrix}, \tag{1.49}$$

$$B_1' = \gamma_1 \beta^{-1} \begin{pmatrix} -(\xi + \mu)\gamma_0 f^* & (\xi + \mu)^2 \\ \gamma_0^2 (f^*)^2 & -(\xi + \mu)\gamma_0 f^* \end{pmatrix}, \tag{1.50}$$

$$C_1' = \gamma_1 \beta^{-1} \begin{pmatrix} -(\xi + \mu)\gamma_0 f & \gamma_0^2 f^2 \\ (\xi + \mu)^2 & -(\xi + \mu)\gamma_0 f \end{pmatrix}, \tag{1.51}$$

and

$$D_1' = \beta^{-1} \begin{pmatrix} (\xi + \mu)(\alpha - \gamma_1^2) & -\gamma_0 f \alpha \\ -\gamma_0 f^* \alpha & (\xi + \mu)\alpha \end{pmatrix}. \tag{1.52}$$

### 1.3.3. Green's function in the real space basis

**Transformation to the real space basis**

Now, we have obtained the Green's function operator $G_0(\epsilon)$ as

$$G_0(\epsilon) = \sum_k C^\dagger(\mathbf{k}) G_0(\epsilon, \mathbf{k}) C(\mathbf{k}) = \sum_k C^\dagger(\mathbf{k})[\epsilon - \mathcal{H}(\mathbf{k})]^{-1} C(\mathbf{k}). \tag{1.53}$$

To express it in the real space basis, one needs to apply the transformation rules

$$C(\mathbf{k}) = \frac{1}{\sqrt{N}} \sum_R \exp(i\mathbf{k} \cdot \mathbf{R}) M(\mathbf{k}) C(\mathbf{R}) \tag{1.54}$$

and

$$C^\dagger(\mathbf{k}) = \frac{1}{\sqrt{N}} \sum_R \exp(-i\mathbf{k} \cdot \mathbf{R}) M^\dagger(\mathbf{k}) C^\dagger(\mathbf{R})$$

$$= \frac{1}{\sqrt{N}} \sum_R \exp(-i\mathbf{k} \cdot \mathbf{R}) C^\dagger(\mathbf{R}) M^\dagger(\mathbf{k}) \tag{1.55}$$

for the creation and annihilation operators, cf. Eq. (1.18), where

$$M(k) = \begin{pmatrix} m(k) & 0 \\ 0 & m(k) \end{pmatrix}, \quad m(k) = \mathrm{diag}(1, \exp(ik \cdot \delta_1), \exp(-ik \cdot \delta_1), 1), \quad (1.56)$$

and

$$C(R) = \big( A_1(R, +), B_1(R + \delta_1, +), A_2(R - \delta_1, +), B_2(R, +),$$
$$A_1^\dagger(R, -), B_1^\dagger(R + \delta_1, -), A_2^\dagger(R - \delta_1, -), B_2^\dagger(R, -) \big)^T. \quad (1.57)$$

Thus, the Green's resolvent can be written as

$$G_0(\epsilon) = \sum_k C^\dagger(k) G_0(\epsilon, k) C(k) = \sum_{R,R'} C^\dagger(R) \tilde{G}_0(\epsilon, R, R') C(R') \quad (1.58)$$

with

$$\tilde{G}_0(\epsilon, R, R') = \frac{1}{N} \sum_k \exp(-ik \cdot (R - R')) \, M^\dagger(k) G_0(\epsilon, k) M(k). \quad (1.59)$$

Note that the blocks $A$, $B$, $C$, and $D$ of $G_0(\epsilon, k)$ depend on $k$ only via $g_0(\xi, k)$, and that the matrix $M(k)$ is block diagonal. Therefore, it is sufficient to calculate

$$\tilde{g}_0(\xi, R, R') = \frac{1}{N} \sum_k \exp(-ik \cdot (R - R')) \, m^\dagger(k) \, g_0(\xi(\epsilon), k) \, m(k). \quad (1.60)$$

Then, Eqs. (1.37) to (1.40) give the following expression for $\tilde{G}_0(\epsilon, R, R')$:

$$\tilde{G}_0(\epsilon, R, R') = \begin{pmatrix} \tilde{A}(\epsilon, R, R') & \tilde{B}(\epsilon, R, R') \\ \tilde{C}(\epsilon, R, R') & \tilde{D}(\epsilon, R, R') \end{pmatrix} \quad (1.61)$$

with

$$\tilde{A}(\epsilon, R, R') = \frac{1}{2} \tilde{g}_+(\xi, R, R') + \frac{\epsilon}{\xi} \tilde{g}_-(\xi, R, R'), \quad (1.62)$$

$$\tilde{B}(\epsilon, R, R') = -\frac{\Delta}{2\xi} \tilde{g}_-(\xi, R, R'), \quad (1.63)$$

$$\tilde{C}(\epsilon, R, R') = -\frac{\Delta^*}{2\xi} \tilde{g}_-(\xi, R, R'), \quad (1.64)$$

$$\tilde{D}(\epsilon, R, R') = -\frac{1}{2} \tilde{g}_+(\xi, R, R') + \frac{\epsilon}{\xi} \tilde{g}_-(\xi, R, R'), \quad (1.65)$$

and

$$\tilde{g}_\pm(\xi, R, R') = \tilde{g}_0(\xi, R, R') \pm \tilde{g}_0(-\xi, R, R'). \quad (1.66)$$

**Calculation of the matrix elements of $\tilde{g}_0$**

The calculation of the matrix elements of $\tilde{g}_0$ is rather lengthy. Therefore, only a sketch for one special case will be shown. The most important results, which are used in Section 1.4, are given in Eqs. (1.75) to (1.77). Some other results as well as some mathematical identities, which have been used in this calculations, can be found in Appendices A and B.

For ease of notation, let $z = \xi + \mu$ in the following. For example, Eq. (1.44) now reads

$$\beta = (z^2 - \gamma_0^2|f|^2)^2 - z^2\gamma_1^2 = [z(z - \gamma_1) - \gamma_0^2|f|^2][z(z + \gamma_1) - \gamma_0^2|f|^2]. \tag{1.67}$$

The sum over all wavevectors $\boldsymbol{k}$ can, for large $N$, be approximated by an integral over the first Brillouin zone such that, e.g.,

$$
\begin{aligned}
(\tilde{g}_0(\xi, \boldsymbol{R}, \boldsymbol{R}'))_{11} &= \frac{1}{N} \sum_{\boldsymbol{k}} \exp(i\boldsymbol{k} \cdot (\boldsymbol{R} - \boldsymbol{R}'))(g_0(\xi, \boldsymbol{k}))_{11} \\
&\approx \frac{V_0}{(2\pi)^2} \int_{1BZ} d^2k \, \exp(i\boldsymbol{k} \cdot (\boldsymbol{R} - \boldsymbol{R}'))(g_0(\xi, \boldsymbol{k}))_{11} \\
&= \frac{V_0}{(2\pi)^2} \int_{1BZ} d^2k \, \frac{\exp(i\boldsymbol{k} \cdot (\boldsymbol{R} - \boldsymbol{R}')) \, z(z^2 - \gamma_0^2|f(\boldsymbol{k})|^2)}{(z^2 - \gamma_0^2|f(\boldsymbol{k})|^2 - z\gamma_1)(z^2 - \gamma_0^2|f(\boldsymbol{k})|^2 + z\gamma_1)}, 
\end{aligned}\tag{1.68}
$$

where $V_0 = 3\sqrt{3}a_{cc}^2/2 \approx 5.24 \times 10^{-2}$ nm$^2$ is the volume (area) of the BLG unit cell. The integral representations of all matrix elements are listed in Appendix A.

The dominant contribution of the integrand comes from the region, where the denominator is zero or has a minimum. This happens near the $\boldsymbol{K}$ and $\boldsymbol{K}'$ points. So, one can utilize circular two-valley approximation

$$\int_{1BZ} d^2k \, F(\boldsymbol{k}) \approx \int_{|q|<\kappa} d^2q \, F(\boldsymbol{K} + \boldsymbol{q}) + \int_{|q|<\kappa} d^2q \, F(-\boldsymbol{K} + \boldsymbol{q}) \tag{1.69}$$

with the momentum cut-off $\kappa = 2(\sqrt{3}\pi)^{1/2}/3a_{cc} \approx 1.10 \times 10^{10}$ nm$^{-1}$ such that the number of states is preserved [compare, e.g., 90]. In this region, $f$ can be approximated by

$$f(\pm\boldsymbol{K} + \boldsymbol{q}) \approx \mp\frac{3}{2}a_{cc}|\boldsymbol{q}| \exp(\pm i\varphi_q), \tag{1.70}$$

see Appendix B, where $\boldsymbol{K}$ is the position of the $\boldsymbol{K}$ point in $\boldsymbol{k}$-space and $|\boldsymbol{q}|$ and $\varphi_q$ are polar coordinates in the vicinity of the $K$ or $K'$ points, respectively. For example,

$$(\tilde{g}_0(\xi, R, R'))_{11} \approx \frac{V_0 v^2}{(2\pi)^2} \int_{|q|<\kappa} d^2q \, \frac{\exp(i(K+q)\cdot(R-R')) \, z(\tilde{z}_0^2 - |q|^2)}{(\tilde{z}_-^2 - |q|^2)(\tilde{z}_+^2 - |q|^2)}$$
$$+ \frac{V_0 v^2}{(2\pi)^2} \int_{|q|<\kappa} d^2q \, \frac{\exp(i(-K+q)\cdot(R-R')) \, z(\tilde{z}_0^2 - |q|^2)}{(\tilde{z}_-^2 - |q|^2)(\tilde{z}_+^2 - |q|^2)}, \quad (1.71)$$

where

$$v = \frac{2}{3a_{cc}\gamma_0}, \quad \tilde{z}_0^2 = v^2 z^2, \quad \text{and} \quad \tilde{z}_\pm^2 = v^2 z_\pm^2 = v^2 z(z \pm \gamma_1). \quad (1.72)$$

In the case $R = R'$, the integration over the polar angle $\varphi_q$ contributes to a factor of $2\pi$ such that, together with the partial fraction decompositions Eqs. (B.2) and (B.3), we have

$$(\tilde{g}_0(\xi, R, R))_{11} = \frac{V_0 v^2 z}{2\pi} \left( \int_0^\kappa dq \, \frac{q}{\tilde{z}_-^2 - q^2} + \int_0^\kappa dq \, \frac{q}{\tilde{z}_+^2 - q^2} \right). \quad (1.73)$$

The indefinite integral in this equation can be solved analytically:

$$I_1(z, q) = \int dq \, \frac{q}{z^2 - q^2} = -\frac{1}{4} \ln\left( [\text{Re}(z^2) - q^2]^2 + [\text{Im}(z^2)]^2 \right)$$
$$+ \frac{1}{2} i \arctan\left( \frac{\text{Re}(z) - q}{\text{Im}(z)} \right) + \frac{1}{2} i \arctan\left( \frac{\text{Re}(z) + q}{\text{Im}(z)} \right), \quad (1.74)$$

see Appendix B. Thus, the diagonal matrix elements of the onsite Green's function $\tilde{g}_0$ are

$$(\tilde{g}_0(\xi, R, R))_{11} = \frac{V_0 v^2}{2\pi} z[I_1(\tilde{z}_-, \kappa) - I_1(\tilde{z}_-, 0) + I_1(\tilde{z}_+, \kappa) - I_1(\tilde{z}_+, 0)] \quad (1.75)$$

and

$$(\tilde{g}_0(\xi, R, R))_{22} = \frac{V_0 v^2}{2\pi} \{[z - \gamma_1][I_1(\tilde{z}_-, \kappa) - I_1(\tilde{z}_-, 0)] + [z + \gamma_1][I_1(\tilde{z}_+, \kappa) - I_1(\tilde{z}_+, 0)]\} \quad (1.76)$$

as well as

$$(\tilde{g}_0(\xi, R, R))_{33} = (\tilde{g}_0(\xi, R, R))_{22} \quad \text{and} \quad (\tilde{g}_0(\xi, R, R))_{44} = (\tilde{g}_0(\xi, R, R))_{11} \quad (1.77)$$

because of the symmetry between the two graphene layers.

## 1.4. Yu-Shiba-Rusinov states

### 1.4.1. Model of the full system

The Hamiltonian of the SBLG with the adatom chemisorbed on the upper graphene layer is given by

$$H = H_{SBLG} + V_{AD}, \tag{1.78}$$

where $H_{SBLG}$ is the Hamiltonian (1.12) of SBLG and $V_{AD}$ describes the adatom and its interaction with the upper graphene layer. In the following, $X = B$ ($X = A$) is used when the adatom is chemisorbed onto a dimer (nondimer) site, i.e., onto a $B_2$ ($A_2$) atom. The position of the adatom is denoted by $R_{AD}$.

Since the adatom is assumed to have a permanent magnetic moment, the degrees of freedom are extended by its spin degrees of freedom denoted by the quantum number $\varsigma$. Then, the annihilation and creation operators of the adatom are denoted by $D(\sigma, \varsigma)$ and $D^\dagger(\sigma, \varsigma)$. The creation and annihilation operators of the carbon atoms now read $X_i(R, \sigma, \varsigma)$ with $X_i(R, \sigma, +) \equiv X_i(R, \sigma) \equiv X_i(R, \sigma, -)$ (the operator $X_i(R, \sigma)$ being the operator from the system without any adatom).

The adatom potential $V_{AD}$ consists of two parts,

$$V_{AD} = V_o + V_s, \tag{1.79}$$

with $V_o$ and $V_s$ describing orbital and spin effects, respectively. The orbital part, where we additionally included the superconducting pairing, reads

$$
\begin{aligned}
V_o = &\sum_{\sigma,\varsigma} (\varepsilon - \mu) D^\dagger(\sigma, \varsigma) D(\sigma, \varsigma) \\
&+ \sum_{\sigma,\varsigma} \omega \left( D^\dagger(\sigma, \varsigma) X_2(R_{AD}, \sigma, \varsigma) + X_2^\dagger(R_{AD}, \sigma, \varsigma) D(\sigma, \varsigma) \right) \\
&+ \sum_{\varsigma} \left( \Delta D^\dagger(-, \varsigma) D^\dagger(+, \varsigma) + \Delta^* D(+, \varsigma) D(-, \varsigma) \right),
\end{aligned}
\tag{1.80}
$$

where $\varepsilon$ describes the onsite potential of the adatom and $\omega$ the coupling between the adatom and the nearest carbon atom. The spin part reads

$$V_s = -J s \cdot S, \tag{1.81}$$

where $s$ is the spin operator of an itinerant electron and $S$ the spin operator of the permanent magnetic moment. The components of the first one are given by

$s_i = \sum_{\sigma,\sigma',\varsigma,\varsigma'} D^\dagger(\sigma,\varsigma)(\sigma_i)_{\sigma,\sigma'} D(\sigma',\varsigma')$. The parameters of the adatom Hamiltonian, Eqs. (1.79) to (1.81), are taken for a hydrogen atom from fitting to density functional theory calculations [91, 92] and read $\varepsilon = 0.25$ eV, $J = -0.4$ eV, and $\omega = 6.5$ eV in the dimer and $\varepsilon = 0.35$ eV, $J = -0.4$ eV, and $\omega = 5.5$ eV in the nondimer case.

The full tight binding model including the hoppings is depicted in Fig. 1.3 on Page 9 for the dimer case.

## 1.4.2. Connection between the Yu-Shiba-Rusinov states and the Green's function

The Green's resolvent of the Hamiltonian (1.78) is given by $G(\epsilon) = (\epsilon - H)^{-1}$. It can be expressed by the Green's resolvent $G_0$ of SBLG via

$$\begin{aligned}
G(\epsilon) &= (\epsilon - H_{\mathrm{SBLG}} - V_{\mathrm{AD}})^{-1} \\
&= \{[\epsilon - H_{\mathrm{SBLG}}][1 - (\epsilon - H_{\mathrm{SBLG}})^{-1} V_{\mathrm{AD}}]\}^{-1} \\
&= (1 - G_0(\epsilon) V_{\mathrm{AD}})^{-1} G_0(\epsilon) \\
&= G_0(\epsilon) + G_0(\epsilon) V_{\mathrm{AD}} G_0(\epsilon) + \dots,
\end{aligned} \tag{1.82}$$

where we used the Neumann series $(1-T)^{-1} = \sum_n T^n$ in the last step to get the Dyson equation. The energies of the YSR states are eigenenergies of $H$ so that the resolvent $G(\epsilon) = (\epsilon - H)^{-1}$ behaves singularly at these energies. Since we do not expect them to coincide with the eigenvalues of $G_0(\epsilon)$, we find them as the energies, where $1 - G_0(\epsilon) V_{\mathrm{AD}}$ has a nontrivial kernel. Since $1 - G_0(\epsilon) V_{\mathrm{AD}}$ is the identity operator in the kernel of $V_{\mathrm{AD}}$, no element of the kernel of $V_{\mathrm{AD}}$ can be in the kernel of $1 - G_0(\epsilon) V_{\mathrm{AD}}$. Thus, it is sufficient to look at the subspace spanned by $X_2^\dagger(\boldsymbol{R}_{\mathrm{AD}}, \sigma, \varsigma)$ and $D^\dagger(\sigma, \varsigma)$. As this subspace is of finite dimension, one gets the following condition for the energies:

$$\det(1 - G_0(\epsilon) V_{\mathrm{AD}}) = 0. \tag{1.83}$$

Note that this condition is equivalent to $\det(1 - V_{\mathrm{AD}} G_0(\epsilon)) = 0$.

This result is in agreement with the $T$ matrix approximation [32]: The $T$ matrix is defined as $T(\epsilon) = V_{\mathrm{AD}}(1 - G_0(\epsilon) V_{\mathrm{AD}})^{-1}$ such that $G(\epsilon) = G_0(\epsilon) + G_0(\epsilon) T(\epsilon) G_0(\epsilon)$; the energies of the YSR states are given by the poles of the $T$ matrix and correspond to the poles of $(1 - G_0(\epsilon) V_{\mathrm{AD}})^{-1}$ [see also 24].

## 1.4.3. Downfolding of the adatom potential

To solve Eq. (1.83), we need to either extend the Green's function calculated in the previous Section in the space spanned by $X_2^\dagger(\boldsymbol{R}_{\mathrm{AD}}, \sigma, \varsigma)$ and $D^\dagger(\sigma, \varsigma)$, or downfold the operator $V_{AD}$ to the space spanned by $X_2^\dagger(\boldsymbol{R}_{AD}, \sigma, \varsigma)$. Here, we choose the second method.

First, let us explain the method of downfolding with a toy model. Assume a system with two subspaces, where the Hamiltonian is given in the following form:

$$H = \begin{pmatrix} H_1 & V \\ V^\dagger & H_2 \end{pmatrix}. \tag{1.84}$$

Then, the eigenequation reads

$$H_1 \Psi_1 + V \Psi_2 = E\Psi_1 \quad \text{and} \quad V^\dagger \Psi_1 + H_2 \Psi_2 = E\Psi_2. \tag{1.85}$$

From the second equation, one has $\Psi_2 = (E - H_2)^{-1} V^\dagger \Psi_1$ which can be inserted into the first equation yielding

$$H_1 \Psi_1 + V(E - H_2)^{-1} V^\dagger \Psi_1 = E\Psi_1. \tag{1.86}$$

This is another eigenvalue-like equation $H' \Psi_1 = E\Psi_1$ with the "downfolded Hamiltonian"

$$H' = H_1 + V(E - H_2)^{-1} V^\dagger. \tag{1.87}$$

The second part, $V(E - H_2)^{-1} V^\dagger$, looks like an additional potential added to the Hamiltonian $H_1$ and is known as self-energy.

In our case, $H_2$ is given by the onsite terms of $V_{AD}$,

$$\sum_{\sigma,\varsigma}(\varepsilon - \mu) D^\dagger(\sigma,\varsigma)D(\sigma,\varsigma) + \sum_\varsigma \left(\Delta D^\dagger(-,\varsigma)D^\dagger(+,\varsigma) + \Delta^* D(+,\varsigma)D(-,\varsigma)\right) - J\boldsymbol{s}\cdot\boldsymbol{S}, \tag{1.88}$$

and $V$ by the hoppings of $V_{AD}$ between the adatom and the graphene layer,

$$\sum_{\sigma,\varsigma} \omega X_2^\dagger(\boldsymbol{R}_{AD}, \sigma, \varsigma)D(\sigma,\varsigma). \tag{1.89}$$

Utilizing once more the anticommutation relations (1.28), these can be rewritten into

$$\frac{1}{2}\sum_{\sigma,\varsigma}(\varepsilon - \mu)\left(D^\dagger(\sigma,\varsigma)D(\sigma,\varsigma) - D(\sigma,\varsigma)D^\dagger(\sigma,\varsigma) + 1\right)$$

$$+ \frac{1}{2}\sum_\varsigma\left(\Delta D^\dagger(-,\varsigma)D^\dagger(+,\varsigma) + \Delta^* D(+,\varsigma)D(-,\varsigma)\right)$$

$$- \frac{1}{2}\sum_\varsigma\left(\Delta D^\dagger(+,\varsigma)D^\dagger(-,\varsigma) + \Delta^* D(-,\varsigma)D(+,\varsigma)\right) - J\boldsymbol{s}\cdot\boldsymbol{S} \tag{1.90}$$

(where the constant $2(\varepsilon - \mu)$, only contributing to the ground state energy, can be ignored) and

$$\frac{1}{2} \sum_{\sigma,\varsigma} \omega\Big( X_2^{\dagger}(R_{AD}, \sigma, \varsigma)D(\sigma, \varsigma) - D(\sigma, \varsigma)X_2^{\dagger}(R_{AD}, \sigma, \varsigma) \Big). \tag{1.91}$$

### 1.4.4. Matrix representations

In the two subspaces, the extended Nambu basis can be chosen as

$$C = \big( X_2(R_{AD}, +, +), X_2(R_{AD}, -, +), X_2(R_{AD}, +, -), X_2(R_{AD}, -, -),$$
$$X_2^{\dagger}(R_{AD}, +, +), X_2^{\dagger}(R_{AD}, -, +), X_2^{\dagger}(R_{AD}, +, -), X_2^{\dagger}(R_{AD}, -, -) \big) \tag{1.92}$$

and

$$C' = \big( D(+, +), D(-, +), D(+, -), D(-, -),$$
$$D^{\dagger}(+, +), D^{\dagger}(-, +), D^{\dagger}(+, -), D^{\dagger}(-, -) \big). \tag{1.93}$$

Let $\mathcal{H}_2, \mathcal{V}, \dots$ be the matrix representations of the Bogoliubov-de Gennes Hamiltonians of $H_2, V, \dots$ in this Nambu space. Using $\zeta = \varepsilon - \mu$, they read

$$\mathcal{H}_2 = \begin{pmatrix} \zeta - J & 0 & 0 & 0 & 0 & -\Delta & 0 & 0 \\ 0 & \zeta + J & -2J & 0 & \Delta & 0 & 0 & 0 \\ 0 & -2J & \zeta + J & 0 & 0 & 0 & 0 & -\Delta \\ 0 & 0 & 0 & \zeta - J & 0 & 0 & \Delta & 0 \\ 0 & \Delta^* & 0 & 0 & -\zeta + J & 0 & 0 & 0 \\ -\Delta^* & 0 & 0 & 0 & 0 & -\zeta - J & 2J & 0 \\ 0 & 0 & 0 & \Delta^* & 0 & 2J & -\zeta - J & 0 \\ 0 & 0 & -\Delta^* & 0 & 0 & 0 & 0 & -\zeta + J \end{pmatrix} \tag{1.94}$$

and

$$\mathcal{V} = \begin{pmatrix} \omega & 0 & 0 & 0 & 0 & 0 & 0 & 0 \\ 0 & \omega & 0 & 0 & 0 & 0 & 0 & 0 \\ 0 & 0 & \omega & 0 & 0 & 0 & 0 & 0 \\ 0 & 0 & 0 & \omega & 0 & 0 & 0 & 0 \\ 0 & 0 & 0 & 0 & -\omega & 0 & 0 & 0 \\ 0 & 0 & 0 & 0 & 0 & -\omega & 0 & 0 \\ 0 & 0 & 0 & 0 & 0 & 0 & -\omega & 0 \\ 0 & 0 & 0 & 0 & 0 & 0 & 0 & -\omega \end{pmatrix}. \tag{1.95}$$

Furthermore, the Nambu-space representation $\mathcal{G}_0(\epsilon)$ of $G_0(\epsilon)$ reads

$$
\mathcal{G}_0 = \begin{pmatrix}
\mathcal{G}'_{11} & 0 & 0 & 0 & 0 & \mathcal{G}'_{12} & 0 & 0 \\
0 & \mathcal{G}'_{11} & 0 & 0 & -\mathcal{G}'_{12} & 0 & 0 & 0 \\
0 & 0 & \mathcal{G}'_{11} & 0 & 0 & 0 & 0 & \mathcal{G}'_{12} \\
0 & 0 & 0 & \mathcal{G}'_{11} & 0 & 0 & -\mathcal{G}'_{12} & 0 \\
0 & -\mathcal{G}'_{21} & 0 & 0 & \mathcal{G}'_{22} & 0 & 0 & 0 \\
\mathcal{G}'_{21} & 0 & 0 & 0 & 0 & \mathcal{G}'_{22} & 0 & 0 \\
0 & 0 & 0 & -\mathcal{G}'_{21} & 0 & 0 & \mathcal{G}'_{22} & 0 \\
0 & 0 & \mathcal{G}'_{21} & 0 & 0 & 0 & 0 & \mathcal{G}'_{22}
\end{pmatrix},
\tag{1.96}
$$

where

$$
\mathcal{G}' = \begin{pmatrix}
(\tilde{G}_0(\epsilon, \mathbf{R}, \mathbf{R}'))_{44} & (\tilde{G}_0(\epsilon, \mathbf{R}, \mathbf{R}'))_{48} \\
(\tilde{G}_0(\epsilon, \mathbf{R}, \mathbf{R}'))_{84} & (\tilde{G}_0(\epsilon, \mathbf{R}, \mathbf{R}'))_{88}
\end{pmatrix}
\tag{1.97}
$$

in the dimer $(X = B)$ and

$$
\mathcal{G}' = \begin{pmatrix}
(\tilde{G}_0(\epsilon, \mathbf{R}, \mathbf{R}'))_{33} & (\tilde{G}_0(\epsilon, \mathbf{R}, \mathbf{R}'))_{37} \\
(\tilde{G}_0(\epsilon, \mathbf{R}, \mathbf{R}'))_{73} & (\tilde{G}_0(\epsilon, \mathbf{R}, \mathbf{R}'))_{77}
\end{pmatrix}
\tag{1.98}
$$

in the nondimer $(X = A)$ case; see Eq. (1.61) for the definition of $\tilde{G}_0(\epsilon, \mathbf{R}, \mathbf{R}')$.

### 1.4.5. Results

The condition (1.83) for the YSR state spectrum cannot be solved analytically. Therefore, the energies of the YSR states are calculated numerically using the NumPy [93, 94] and SciPy [95, 96] libraries. However, it turned out that it is numerically more stable to determine the roots

$$
\det(\mathcal{V}(E - \mathcal{H}_2)^{-1}\mathcal{V}^\dagger)\det(1 - \mathcal{G}_0\mathcal{V}(E - \mathcal{H}_2)^{-1}\mathcal{V}^\dagger) = 0.
\tag{1.99}
$$

Note that the matrix representations (1.94) to (1.96) from the previous Section and the "downfolded Hamiltonian" (1.87) have been used.

In Fig. 1.4, the spectrum of the YSR states in SBLG is shown for a chemisorbed hydrogen adatom. Let us first focus on the upper panel with $\Delta = 50\,\text{meV}$ and compare our results with numerical results of M. Barth [24]. The latter ones are obtained by diagonalizing a rectangular flake with hard-wall boundary conditions using GNU Octave [97, for details, see 24]. One can see that the analytical and numerical results match very well. The remaining tiny difference can be attributed to the following

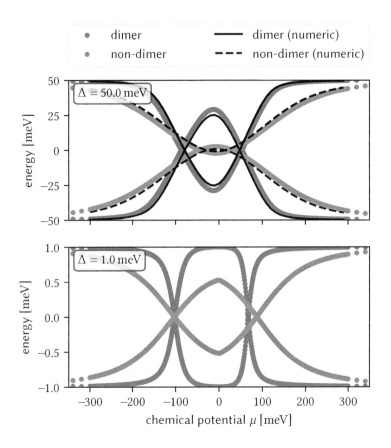

*Figure 1.4. YSR state spectrum of hydrogenated SBLG.* The pairing gap is chosen as $\Delta = 50$ meV in the upper panel and $\Delta = 1$ meV in the lower panel. In the upper panel, the results obtained from Eq. (1.99) (dots) are compared to the numerical results from M. Barth [24] (lines), whereas for the more realistic value of $\Delta = 1$ meV in the lower panel, there are no numerical results because the computational effort is much higher in this case.

reasons: first, the numerics include finite size effects since the sample size cannot be arbitrarily large; and second, the numerical diagonalization has a finite accuracy (here, a tolerance of $10^{-5}$ was used [24]).

Now, consider the more realistic case $\Delta = 1\,\mathrm{meV}$ displayed in the lower panel of Fig. 1.4.[4] We obtain a pair of YSR states in each, dimer and nondimer, case. Note that the difference between the dimer and nondimer case does not only stem from the different values of the parameters $\varepsilon$, $J$, and $\omega$ (see Section 1.4.1, Page 19) but also the different expressions for the diagonal elements of the Green's function (compare, e.g., Section 1.3, Eqs. (1.75) to (1.77)). The YSR "spectra" look similar to the results for a classical spin (see Fig. 1.1) and also agree with the findings of Kochan et al. for hydrogenated single layer graphene [23]. Similar to the two parity switching points with $E(J_cS) = 0$, we also obtain two points with $E(\mu_c) = 0$ with $\mu_c \approx -102\,\mathrm{meV}$ and $\mu_c \approx 67\,\mathrm{meV}$ in the dimer case and $\mu_c \approx -102\,\mathrm{meV}$ and $\mu_c \approx 87\,\mathrm{meV}$ in the nondimer case. However, the YSR states do not peak at $\pm\Delta$ in the nondimer case (this also occurs for $\Delta_0 = 50\,\mathrm{meV}$ in both, dimer and nondimer, cases). This happens because we investigate the $\mu$-dependence of the YSR states instead of the $J$-dependence.

Let us now investigate the $J$-depencence of the YSR states by treating $J$ as a free variable[5]. The results are shown in Fig. 1.5 for different values of the chemical potential $\mu$. We again observe the characteristic YSR state behaviour: There exists a pair of YSR states which behaves qualitatively like the results of Shiba [13]—especially, there are two parity switching points $J_c$ (where $E(J_c) = 0$) and there are no subgap sates for $J = 0$ (i.e., $E(J = 0) = \pm\Delta$). Previously, Lado and Fernández-Rossier [85] reported an unconventional linear dependence of the energy on the exchange coupling $J$ around $J = 0$ in hydrogenated superconducting single layer graphene. However, our results in Fig. 1.5 for SBLG clearly display a conventional quadratic behaviour like in Fig. 1.1. There are several possible explanations for this difference: The different number of graphene layers, the different models for the adatom or a different choice of parameters. Applying our model to single layer graphene would lead to very similar results as in Sections 1.4.1 to 1.4.4, the major difference being different Green's function elements in Eq. (1.96), directly related to $\tilde{g}_0(\xi, \boldsymbol{R}, \boldsymbol{R}')$. Moreover, we think that it is unlikely that a different choice of the parameters (namely, the superconducting coupling $\Delta$, the adatom onsite potential $\epsilon$, and the coupling $\omega$ between the adatom and the carbon atom) can affect the qualitative behaviour of the YSR states when sticking

---

[4]Note that there are no numerical results for $\Delta = 1\,\mathrm{meV}$ because the computational effort is too high.

[5]In experiment, it is hard to change the exchange coupling $J$ while it is much more feasible to tune the chemical potential $\mu$ via gating.

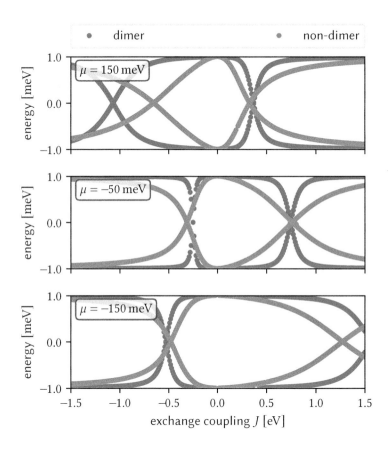

*Figure 1.5. Dependence of the YSR states in hydrogenated SBLG on the exchange coupling J. For all values of the chemical potentials $\mu$, the pairing gap is set to $\Delta = 1$ meV.*

to realistic magnitudes. On the other hand, instead of including an additional atom in the tight-binding Hamiltonian (compare Eqs. (1.12), (1.78) and (1.79)), Lado and Fernández-Rossier add an infinite onsite energy on the functionalized carbon atom. We think that this infinity is the reason for the nonanalytic behaviour of their spectral function and, thus, also the reason for the unconventional behaviour of the YSR states they observed.[6]

**Outlook**

We have shown that YSR states exist in SLBG and show the characteristic dependence on $\mu$ and $J$. This motivates experimental investigation of these systems. Additionally, YSR states in graphene systems with adatoms have been reported to modify the spin relaxation rates [23, 24]: Normally, the spin relaxation decreases when increasing the temperature in the presence of magnetic impurities but increases in the presence of spin-orbit coupling (all in the superconducting phase); the first phenomenon is known as Hebel-Slichter effect [98–100]. However, resonant YSR states lead to a breakdown of this effect in both, single [23] and bilayer [24] graphene.

---

[6]Note that a change in the model can, quite generally, have a massive impact on the physical properties of a system.

# Chapter 2.

# Crossed Andreev reflection in topological insulator nanowire T junctions

The results of this chapter have been published by Fuchs et al. [101]. The 2D surface model has been developed by Jacob Fuchs, who also did the numerical calculations based upon it; the results obtained with the 3D BHZ model have been obtained by Michael Barth.

## 2.1. Introduction

### Crossed Andreev reflection

Superconductor (S) hybrid structures show many interesting physical effects [4] and there is still thriving research in this field since the first observation of super-conductivity in 1911 by Kamerlingh Onnes [4, 102]. One particularly interesting phenomenon in NS junctions—a junction between a normal conductor (N) and a superconductor—is the Andreev reflection [6, 103]: Here, an electron incoming from the N contact with an energy inside the superconducting band gap is reflected as a hole, whereas simultaneously a Cooper pair is formed in the S. In an NSN junction, where a S is sandwiched in between two N contacts, there exist more possible scattering processes, all of them beeing depicted in Fig. 2.1. Of particular interest is the so-called crossed Andreev reflection (CAR), where the outgoing hole is located in the second N contact such that the created Cooper pair is formed from two electrons from different, spatially separated N contacts. Its reciprocal process, Cooper pair "splitting", where two entangled electrons in separated leads are generated, is of interest as a possible source of entangled electrons [26, 27]. Cooper pair splitters have already been investigated both theoretically [25, 104, 105] and experimentally [26, 27,

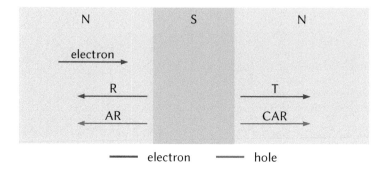

*Figure 2.1. Possible scattering processes in an NSN junction.* Reflection (R): An electron (red) is reflected as an electron. Transmission (T): An electron is transmitted as an electron. Andreev reflection (AR): An electron is reflected as a hole (blue). Crossed Andreev reflection (CAR): An electron is transmitted as a hole.

106]. Also, CAR has been reported in experiments [107] and theoretically predicted to exist, e.g., in the 1D Kitaev chain [108] or in graphene [109–111]. Although CAR is in most cases dominated by electron transmission (T), where the incoming electron leaves the junction through the other N contact as an electron, perfect CAR has been reported in bilayer graphene [112], Dirac semimetals in the quantum Hall regime [113], or 2D antiferromagnets [114], for example. However, these findings depend strongly on geometrical properties or on local doping.

### Topological insulators and 3D TI nanowires

Topological insulators (TIs) are interesting materials exhibiting a gapped, insulating bulk, but gapless surface states capable of transporting charge [3]. This leads to a lot of novel transport phenomena: 2D TIs host 1D helical edge modes [115], whereas 3D TIs possess 2D surface states which mediate the transport robust to disorder [116]. In contact with s-wave superconductors, they show topological superconductivity and can host Majorana zero modes [3, 117]. In this work, we focus on nanowires made from 3D TIs. Such nanowires have been investigated in several experiments [e.g. 118–138].

### Crossed Andreev reflection in topological insulators

Due to the many diverse transport phenomena happening in TIs, we are interested in the question, whether CAR can also be realized in this class of materials. Indeed, CAR has already been discussed previously. For example, it has been used as study tool for topological phase transitions and Majorana zero modes [139–141]. Furthermore, it has been predicted that CAR should occur in systems with magnetic ordering [142, 143], in double TI Josephson junctions [144], in 2D TIs where the two edges are coupled [145, 146], in the presence of odd-frequency triplet superconductivity [147, 148], or in bipolar setups [149–151].

**Outline**

In this Chapter, we propose and investigate the T junction device depicted in Fig. 2.5, an experimentally realizable setup exhibiting CAR. On the one hand, our proposed system shows perfect CAR in a large parameter range (meaning that no T and no reflections occur, which is quite remarkable), but also shows (imperfect) CAR in an even larger range. These findings are robust to disorder due to topological protections and are even controllable by an external magnetic field—meaning that one can switch between CAR and T by tuning the magnetic field.

We want to start our exploration by reviewing the basics of 3D TI nanowires that are necessary to understand the rest of this Chapter as well as the contents of Chapter 3. Afterwards, we describe the 3D TI nanowire NS junction from Juan, Ilan, and Bardarson [152] our proposed T junction device is built upon in Section 2.3. Then, we discuss the working principle of our proposed setup in Section 2.4 and elaborate on the theoretical model in Section 2.5. The numerical results are presented in Section 2.6. In Appendix C, we give some technical details about our numerical implementation.

## 2.2. Introduction to 3D TI nanowires

Nanowires made from 3D TIs have already been extensively investigated in literature both theoretically [e.g. 116, 133, 152–160] and experimentally [e.g. 118–138]. In this Section, we want to give a brief overview of their basic properties which are necessary to understand the rest of this Chapter. There exist good introductions by Kozlovsky [161] and, especially, Bardarson and Ilan [154] who also cover the contents of Section 2.3.

## 2.2.1. 3D TI nanowires with parallel magnetic field

The topological surface states of a cylindrical nanowire are commonly described by the Dirac Hamiltonian [116, 133, 154, 156, 161],

$$H = \hbar v_F \left[ \hat{k}_x \sigma_x + \left( \hat{k}_s + \frac{2\pi}{P} \frac{\phi}{\phi_0} \right) \sigma_y \right] - \mu. \tag{2.1}$$

Here, $v_F$ is the Fermi velocity, $x$ and $s$ are the coordinates along the wire and around the circumference, $\hat{k}_x$ and $\hat{k}_s$ the cooresponding wave number operators, $P$ denotes the perimeter of the cross section, $\phi$ the flux of the magnetic field parallel to the wire, $\phi_0 = h/e$ is the magnetic flux quantum, and $\mu$ the chemical potential. However, there is a Berry phase of $\pi$ around the circumference [116, 155, 156, 162, 163]—therefore, the wave function satisfies antiperiodic boundary conditions, $\psi(s) = -\psi(s + P)$ [for a detailed derivation of the Hamiltonian (2.1) and the boundary condition, see 156, 161, 164]. Due to the antiperiodic boundary conditions, the azimuthal wave number is quantized as $k_s = 2\pi(l + 1/2)/P$ with $l \in \mathbf{Z}$. For a given wave number $k_x$ and a given angular momentum quantum number $l$, there are two eigenmodes with energies

$$E = \pm \hbar v_F \sqrt{k_x^2 + \frac{4\pi^2}{P^2} \left( l + \frac{1}{2} + \frac{\phi}{\phi_0} \right)^2} - \mu. \tag{2.2}$$

Notice that, for $\mu = 0$, the spectrum (2.2) is gapped whenever $1/2 + \phi/\phi_0 \notin \mathbf{Z}$, see Fig. 2.2. For $1/2 + \phi/\phi_0 \in \mathbf{Z}$ on the other hand, there exists a mode with linear dispersion. This mode garanties that the conductance does not drop below one conductance quantum $e^2/h$ in the presence of disorder [116, 154, 159] and is, therefore, called "perfectly transmitted mode".

Before we continue, we have to comment on the geometry. To get Eq. (2.1), a circular cross section has been taken. In experiments however, the cross section is often rectangular or trapezoidal [see, e.g., 118, 133]. Then, one can either still use Hamiltonian (2.1) together with the according perimeter $P$ or use a Hamiltonian which respects the correct surface orientations but needs more intricate matching conditions at the edges [compare 165]. While the first approach is much better suited for analytical insights, we need to use the latter one for our T junction device, see Section 2.5. Thus, we only consider nanowires with rectangular cross sections from now on.

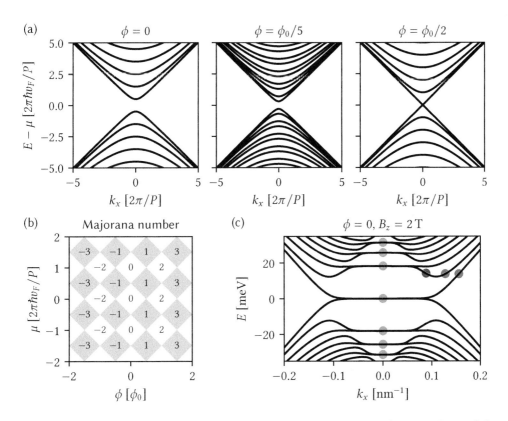

*Figure 2.2. Spectra of the 3D TI nanowires.* (a) Spectra for three different values of the magnetic flux $\phi$ through the wire cross section, where $\phi_0 = h/e$ is the magnetic flux quantum; the chemical potential $\mu$ is set to $\mu = 0$ for these spectra. For $\phi = \phi_0/2$, the gap is closed and a linear mode exists. (b) Phase diagram of the 3D TI nanowires. White regions are topologically trivial with Majorana number $M = +1$, green regions are topologically nontrivial with $M = -1$; the Majorana number $M$ is given as $M = (-1)^\nu$, where $\nu$ is the number of Fermi points. The numbers in the regions indicate the number of Fermi points $\nu$. (c) Spectrum of a 3D TI nanowire with a perpendicular magnetic field; parameters and model are the same as for Fig. 2.4. Landau levels are marked green and propagating (counterpropagating) edge states of one side blue (red).

## 2.2.2. 3D TI nanowires in a perpendicular magnetic field

Let us now switch to the case when there is a magnetic field perpendicular to the wire axis, say, in vertical direction. When the field is strong enough, a quantum Hall transition happens: Landau levels form on the top and bottom surfaces and chiral edge states form on the side surfaces [158, 163, 165–167]. "Strong enough" means that the magnetic length $l_B = \sqrt{\hbar/eB}$, which characterizes the spatial elongation of the Landau level, is smaller than the wire width. In the spectrum, Landau levels manifest as flat bands (marked with green in Fig. 2.2(c)), whereas the bent bands at their edges correspond to the propagating states on the side surfaces (marked with red and blue in Fig. 2.2(c)). Note that for our geometry some bands show dips near the edge of the Landau levels such that two additional side surface states appear before the Fermi level crosses the next Landau level. In this case, one of these modes propagates in the opposite direction (this one is marked with red in Fig. 2.2(c)). Thus, the difference between the number $N_+$ of propagating modes and the number $N_-$ of counterpropagating modes equals $N_+ - N_- = 2n + 1$ [166], where $n$ is the number of filled Landau levels.

## 2.2.3. Superconducting 3D TI nanowires

When in contact to a superconductor, a 3D TI nanowire can become superconducting itself which is known as the proximity effect. Cook and Franz [28] investigated a nanowire proximitized by an *s*-wave superconductor and showed that it exhibits topological superconductivity for suitable values of the chemical potential $\mu$ and the axial magnetic flux $\phi$. This can be explained using the Majorana number $M$. As shown by Kitaev [19], the Majorana number simplifies to $M = (-1)^\nu$ in the limit of small superconducting coupling $\Delta$, where $\nu$ is the number of Fermi points in the positive half of the Brillouin zone. Whenever one has $M = +1$, the system is topologically trivial, whereas it is topologically nontrivial for $M = -1$. For the 3D TI nanowire, the nontrivial regions are colored green in phase diagram Fig. 2.2(b). Note that for $1/2 + \phi/\phi_0 \in \mathbf{Z}$, the number of Fermi points is always odd ensuring the nontrivial topology for any value of the chemical potential $\mu$. Thus, the (proximitized) nanowire represents a topological superconductor in this case. As 1D topological superconductors are predicted to have two boundary Majorana states at each end [19], these nanowires are promising candidates for realizing Majorana zero modes (which can also be tuned using the axial magnetic field). Note that the existence of Majorana modes in exactly these systems has been verified both analytically [28] and numerically [28, 29].

Let us now discuss how the Hamiltonian of the superconducting nanowire looks like. The Bogoliubov-de Gennes Hamiltonian of the system reads

$$H_{\mathrm{BdG}} = \begin{pmatrix} H(\phi) & \Delta_0 \exp(2\pi i n_v s/P) \\ \Delta_0 \exp(-2\pi i n_v s/P) & -H(-\phi) \end{pmatrix} \qquad (2.3)$$

with $H$ from Eq. (2.1) [152, 154, 157, 164]. Here, we have allowed for the existence of vortices along the nanowire axis by including the phase $2n_v s/P$ in the superconducting pair potential; $n_v$ describes the number of vortices. When the magnetic flux is 0 ($\phi_0/2$; $\phi_0$; etc.), it is expected to have $n_v = 0$ (1; 2; etc.) vortices pinned, the pinning of the new vortices happening somewhere in between (e.g. $n_v = \mathrm{floor}(2\phi/\phi_0 + 1/2)$ whith the floor function $\mathrm{floor}(x) = \max\{n \in \mathbb{Z} | n \leq x\}$). This is crucial to maintain the superconducting gap, otherwise the magnetic field closes the gap [this is extensively discussed in 29, 164, see also 152, 154].[1]

## 2.3. 3D TI nanowire NS junction

Before presenting the T junction, we have to introduce its most important building block, the 3D TI nanowire NS junction shown in Fig. 2.3. This system has been investigated by Juan, Ilan, and Bardarson [152, 154].

In a particle-hole symmetric system, the eigenvalues of the reflection matrix are either twofold degenerate or equal to 0 or 1 [a selfcontained derivation is given in 169]. This is known as Béri degeneracy for it was discovered by Béri [170]. When there is only one conducting mode present in the N part, the reflection coefficient can only take the values 0 or 1. Thus, the NS conductance $G_{\mathrm{NS}}$ is either 0 or $2e^2/h$. It now happens that the conductance is 0 in the topologically trivial regime and $2e^2/h$ in the topologically nontrivial regime [152, 169]. This can be clearly seen in the NS conductances shown in Fig. 2.4. Note that, in a clean system without perpendicular magnetic field, the conductance shows plateaus at even (odd) integer multiples of $2e^2/h$ in the topologically trivial (nontrivial) case when more than one mode is present [21, 169]. In the presence of disorder, the plateaus of the single-mode regime remain [152, 169] since the Béri degeneracy still applies.

---

[1]The phase $2n_v s/P$ of the pair potential is similar to the one of the Hamiltonian (3.1) in Chapter 3 so that its derivation in Appendix D is also applicable to the system in this Chapter. However, we have to restrict to integer $n_v$ since the superconducting pair potential $\Delta(s) = \Delta_0 \exp(2i n_v s/P)$ has to be periodic in $s$: $\Delta(s + P) = \Delta(s)$.

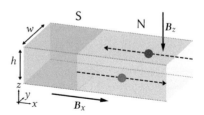

*Figure 2.3. Scheme of the 3D TI nanowire NS junction* from Juan, Ilan, and Bardarson [152]. One half of the 3D TI nanowire is in the normal state (N; drawn in gray) and the other half is superconducting (S) through the proximity effect (drawn in green). The parallel magnetic field $B_x$ is present in the whole junction, whereas the perpendicular magnetic field $B_z$ is only present in the N part and does not extend into the S part.

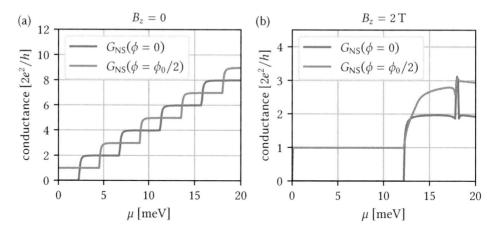

*Figure 2.4. Conductance of the 3D TI nanowire NS junction* from Fig. 2.3 without (a) and with (b) a perpendicular magnetic field in the N part. Calculations are based on the model from Juan, Ilan, and Bardarson [152] and were performed using the Kwant code [168]. The following parameters have been used: $\hbar v_F = 330$ meV nm, $\Delta_0 = 0.25$ meV, $w = 160$ nm, and $h = 70$ nm.

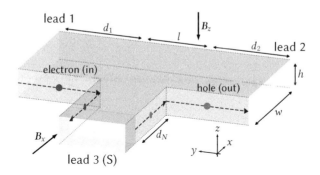

*Figure 2.5. Scheme of the T junction.* The setup consists of three TI nanowires, where the third one (lead 3) is proximitized by an s-wave superconductor. The magnetic field $B_z$ is perpendicular to the junction and induces chiral edge states in the TI nanowires. Thus, electrons entering the junction from lead 1 have to leave the junction through lead 2 either as electron or hole. These processes are the electron transmission (T) and crossed Andreev reflection (CAR), respectively. The magnetic field $B_x$ parallel to the third lead allows to control which of these processes occurs.

The single-mode regime is the most interesting for our purpose: $G_{NS} = 0$ in the trivial regime ($\phi = 0$) indicates T, whereas $G_{NS} = 2e^2/h$ in the nontrivial regime ($\phi = \phi_0/2$) indicates AR. This means that one can switch between T and AR by tuning the magnetic field. When there is no perpendicular magnetic field ($B_z = 0$), the range of the single-model regime is relatively small: the chemical potential $\mu$ has to be tuned into a range determined by $\hbar v_F 4\pi/P$. However, we can enlarge this regime considerably by adding a perpendicular magnetic field in the N part ($B_z \neq 0$) to drive it into the quantum Hall regime: In this case, the parameter range for the chemical potential $\mu$ is determined by the position of the first Landau level $\hbar v_F \sqrt{2e|B_z|/\hbar} = \hbar v_F \sqrt{2/l_{B_z}}$.

## 2.4. T junction

In the NS junction from the previous Section, the incoming electron and the outgoing hole are already spatially separated into the two chiral edge states (for $\phi = \phi_0/2$). We take advantage of this by splitting these chiral modes into different leads converting AR into CAR. Figure 2.5 shows the T junction device proposed by us, where this can be achieved.

Let us take a look at the transport process depicted in Fig. 2.5 to illustrate the working principle of the T junction device: First, note that the magnetic field $B_z$ is perpendicular to all three leads such that all of them are in the quantum Hall regime; this is crucial for our purposes. (In the superconducting part, the perpendicular magnetic field is screened, as above.) Thus, an electron entering the junction through lead 1 through the chiral front surface channel cannot go to the top and bottom surfaces because they are insulating due to the quantum Hall effect. Therefore, it enters the side surface of lead 3, where it eventually bumps into the NS interface. There, it is Andreev reflected (normally reflected) when the magnetic field $B_x$ parallel to the third lead induces a magnetic flux $\phi = \phi_0/2$ ($\phi = 0$) through the cross section. The outgoing hole (electron) on the other chiral channel enters the front surface channel of lead 2 for the same reasons mentioned before leaving the junction through lead 2. In total, the electron from lead 1 leaves the junction through lead 2 as hole (electron) such that we have CAR (T).

It is important to notice that the T junction behaves differently depending on the magnitude of the magnetic field $B_x$. This means that one can switch between CAR and T by tuning $B_x$.

Note that the parallel field $B_x$ should not be too strong, otherwise Landau levels develop on the front and back surfaces of leads 1 and 3, too. This restriction is, however, easily met since the width of the nanowires is typically much larger than the height in experimental realizations [see, e.g., 133]; we tried to show this in Fig. 3.2.

## 2.5. Methods

The Hamiltonian Eq. (2.1) cannot be applied to the T junction since there are two wires with different axis orientations. Thus, we use an approach similar to the one of Brey and Fertig [165], where each surface is described by a different Hamiltonian and the wave functions are matched at the edges, see Section 2.5.1. From this Hamiltonian, we derive a tight-binding model via the finite difference method as described in Appendix C. Afterwards, the S matrix is calculated using the Kwant code [168, see also 171–173] from which we extract the transmission coefficients as well as the local and nonlocal conductances $G_{11}$ and $G_{21}$, see Section 2.5.2.

## 2.5.1. Model

Before we identify the Hamiltonian of the system, let us have a look at the geometry. As depicted in Fig. 2.5, there is one wire oriented in the $y$ direction and another one oriented in the $-x$ direction. The latter one forms the T arm and becomes superconducting for $x < 0$. Width and height of the first wire are denoted by $w$ and $h$, the width of the second wire is denoted by $l$. The superconducting part is $d_N$ away from the first wire and, when implementing the system, leads 1 and 2 are put $d_1$ and $d_2$ away from the T arm, although the explicit choice of these distances does not matter.

A surface with the normal $\hat{n}$ (which is a unit vector) is described by the Hamiltonian

$$H_{\hat{n}} = \hbar v_F(\boldsymbol{\sigma} \times \hat{\boldsymbol{k}}) \cdot \hat{\boldsymbol{n}} - \mu, \tag{2.4}$$

where $\boldsymbol{\sigma}$ is the vector of Pauli matrices and $\hat{\boldsymbol{k}}$ the vector of wave number operators [156, 165]. For $Bi_2Se_3$, we use $\hbar v_F = 410\,\mathrm{meV\,nm}$ [174], whereas $\hbar v_F = 330\,\mathrm{meV\,nm}$ for HgTe [133]. Note that we neglect any anisotropy and doping effects by using the same value for $\hbar v_F$ for any surface and by omitting any scalar potential. Written out, one has

$$H_{\pm\hat{x}} = \pm\hbar v_F(-\hat{k}_y\sigma_z + \hat{k}_z\sigma_y), \tag{2.5}$$

$$H_{\pm\hat{y}} = \pm\hbar v_F(\hat{k}_x\sigma_z - \hat{k}_z\sigma_x), \tag{2.6}$$

and

$$H_{\pm\hat{z}} = \pm\hbar v_F(-\hat{k}_x\sigma_y + \hat{k}_y\sigma_x). \tag{2.7}$$

For implementing the wire Hamiltonians, it is best to express the wave number operators as one parallel to the wire, $\hat{k}_\parallel$, and one going counterclockwise around the perimeter, $\hat{k}_\perp$, analogously to Eq. (2.1). The resulting Hamiltonians are listed in Table 2.1.

At an edge connecting the $\hat{n}_1$ surface with the $\hat{n}_2$ surface, the wave functions satisfy $\psi_{\hat{n}_1} = U_{\hat{n}_1\hat{n}_2}\psi_{\hat{n}_2}$ with $U_{\hat{n}_1\hat{n}_2} = \exp(-i\boldsymbol{\theta} \cdot \boldsymbol{\sigma}/2)$ beeing the appropriate spin rotation and $\boldsymbol{\theta}$ denoting the product of the rotation angle and axis. These matching conditions fit the one of Brey and Fertig [165] for the isometric parameters $A_1 = A_2 = \hbar v_F$ and $D_1 = D_2 = 0$. For example, consider the edge between the $\hat{z}$ and $\hat{x}$ surfaces: The wave function $\psi_{\hat{x}}$ on the $\hat{x}$ surface is related to the wave function $\psi_{\hat{z}}$ by a rotation of $\pi/2$ around the $y$ axis:

$$\psi_{\hat{x}} = U_{\hat{x}\hat{z}}\psi_{\hat{z}}, \quad \text{where} \quad U_{\hat{x}\hat{z}} = \exp\left(-i\frac{\pi}{4}\sigma_y\right) = \frac{1}{\sqrt{2}}(1 - i\sigma_y). \tag{2.8}$$

*Table 2.1. Surface Hamiltonians for wires in different directions.* For simplicity, $\hbar v_F = 1$ in this table.

| | wire direction | | |
|---|---|---|---|
| surface | $\hat{x}$ | $\hat{y}$ | $\hat{z}$ |
| $-\hat{y}$ | $-\hat{k}_\parallel \sigma_z + \hat{k}_\perp \sigma_x$ | | $\hat{k}_\parallel \sigma_x + \hat{k}_\perp \sigma_z$ |
| $\hat{x}$ | | $-\hat{k}_\parallel \sigma_z + \hat{k}_\perp \sigma_y$ | $\hat{k}_\parallel \sigma_y + \hat{k}_\perp \sigma_z$ |
| $\hat{z}$ | $-\hat{k}_\parallel \sigma_y + \hat{k}_\perp \sigma_x$ | $\hat{k}_\parallel \sigma_x + \hat{k}_\perp \sigma_y$ | |
| $\hat{y}$ | $\hat{k}_\parallel \sigma_z + \hat{k}_\perp \sigma_x$ | | $-\hat{k}_\parallel \sigma_x + \hat{k}_\perp \sigma_z$ |
| $-\hat{x}$ | | $\hat{k}_\parallel \sigma_z + \hat{k}_\perp \sigma_y$ | $-\hat{k}_\parallel \sigma_y + \hat{k}_\perp \sigma_z$ |
| $-\hat{z}$ | $\hat{k}_\parallel \sigma_y + \hat{k}_\perp \sigma_x$ | $-\hat{k}_\parallel \sigma_x + \hat{k}_\perp \sigma_y$ | |

Indeed, the Hamiltonian satisfies

$$U_{\hat{x}\hat{z}} H_{\hat{z}}(k_x = k_{-z}, k_y = k_y) U_{\hat{x}\hat{z}}^\dagger = H_{\hat{x}}(k_x, k_y). \tag{2.9}$$

Note that the antiperiodic boundary conditions belonging to Hamiltonian (2.1) do not apply here anymore: it is "split" into the matching conditions at the four edges since

$$\exp\left(-\mathrm{i}\frac{\pi}{4}\hat{\boldsymbol{n}}\cdot\boldsymbol{\sigma}\right)^4 = \exp(-\mathrm{i}\pi\hat{\boldsymbol{n}}\cdot\boldsymbol{\sigma}) = \cos(\pi) + \mathrm{i}(\hat{\boldsymbol{n}}\cdot\boldsymbol{\sigma})\sin(\pi) = -1. \tag{2.10}$$

In the superconducting half, the perpendicular magnetic field vanishes, $\boldsymbol{B} = B_x\hat{\boldsymbol{e}}_x + B_z\Theta(x)\hat{\boldsymbol{e}}_z$. The vector potential $\boldsymbol{A}$ is chosen as

$$\boldsymbol{A} = \boldsymbol{A}_x + \boldsymbol{A}_z = \frac{1}{2}B_x(-z\hat{\boldsymbol{e}}_y + y\hat{\boldsymbol{e}}_z) - B_z x\Theta(x)\hat{\boldsymbol{e}}_y. \tag{2.11}$$

Note that it is important to keep $\boldsymbol{A}_z$ constant for $x < 0$ and contiuous at $x = 0$, the most practical choice for the implementation beeing $\boldsymbol{A}_z(x < 0) = 0$.

Superconductivity is modelled with the Bogoliubov-de Gennes Hamiltonian (2.3).

### 2.5.2. Local and nonlocal conductances

In a multiterminal setup, the conductance can be calculated from the transmission coefficients $T_{aa}^{R/AR}$ and $T_{ba}^{T/CAR}$, $b \neq a$, and the number of channels $N_a$ of lead $a$ in the way described by Lambert and Raimondi [175]. The local conductance $G_{aa}$ relates the current in lead $a$ flowing *into* the junction to the voltage applied at the same lead $a$ and reads

$$G_{aa} = \left.\frac{\partial I_a}{\partial V_a}\right|_{V_a=0} = \frac{e^2}{h}\left(N_a + T_{aa}^{\text{AR}} - T_{aa}^{\text{R}}\right); \tag{2.12}$$

the nonlocal conductance $G_{ba}$, $b \neq a$, on the other hand, relates the current in lead $b$ flowing *out of* the junction to the voltage applied at the (differing) lead $a$ and reads

$$G_{ba} = -\left.\frac{\partial I_b}{\partial V_a}\right|_{V_a=0} = \frac{e^2}{h}\left(T_{ba}^{\text{T}} - T_{ba}^{\text{CAR}}\right). \tag{2.13}$$

Note that the sign of the nonlocal conductance $G_{ba}$ indicates the direction of current flow in lead $b$. Therefore, it provides a direct measure for the behaviour of the T junction: A positive nonlocal conductance (e.g. $G_{21} > 0$) indicates that T (from lead 1 to lead 2) dominates over CAR, whereas a negative nonlocal conductance ($G_{21} < 0$) signals dominating CAR.

## 2.6. Results

### Results for a Bi$_2$Se$_3$ nanowire

Let us first examine the results for Bi$_2$Se$_3$. To compare with the numerical results obtained with the more realistic 3D BHZ model, we use rather small wire dimensions (3D calculations with larger dimensions are numerically too demanding): width and height are chosen as $w = l = 50$ nm and $h = 10$ nm. While these dimensions are still experimentally feasible [see 135, for example], they impose the need for large magnetic field to ensure that the wires are in the quantum Hall regime: Here, we use a perpendicular magnetic field of $B_z = 20$ T, which ensures that the N part is in the quantum Hall state as $l_{B_z} = \sqrt{\hbar/eB} \approx 5.7$ nm $\ll w, l$. This order of magnitude does not impose big problems for theoretical simulations, but is unrealistic to be achieved in experiments. Furthermore, we use $\hbar v_F = 410$ meV nm [174] and $\Delta = 0.25$ meV [as in 152]. The sign of $B_z$ is fixed in such a way that it fits the situation from Section 2.4 and Fig. 2.5, where CAR is possible for an electron coming from lead 1. Thus, we exclusively look at the situation, where a small bias voltage is applied at lead 1. The results for the transmission coefficients $T_{21}^{\text{T/CAR}}$ and $T_{11}^{\text{R/AR}}$ are shown in Fig. 2.6, whereas the results for the nonlocal conductances $G_{21}$ are shown in Fig. 2.7.

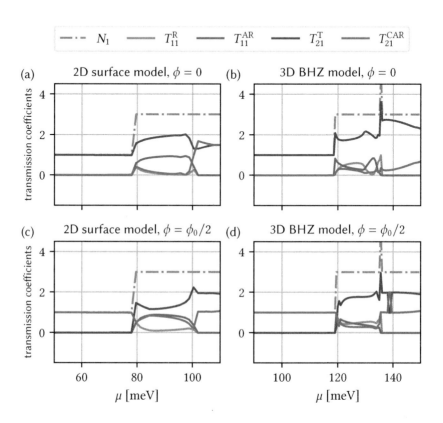

*Figure 2.6. Transmission coefficients in a Bi$_2$Se$_3$ T junction.* Subfigures (a) and (c) show the results obtained with the 2D surface model from Section 2.5, subfigures (b) and (d) the results of Michael Barth from the 3D BHZ model [101]. We used the following parameters: $w = 50$ nm, $h = 10$ nm, $\Delta = 0.25$ meV, and $B_z = 20$ T, together with $\hbar v_F = 410$ meV nm in the 2D surface model calculations. For $\phi = \phi_0/2$, a vortex is present in the superconductor.

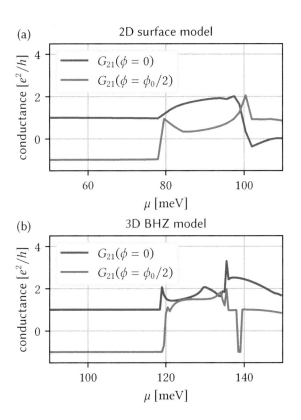

*Figure 2.7. Nonlocal conductance in a Bi$_2$Se$_3$ T junction.* The results are obtained (a) with the 2D surface model from Section 2.5 and (b) with the 3D BHZ model by Michael Barth [101]. The parameters are the same as in Fig. 2.6. In the regions with negative nonlocal conductance, CAR dominates over T.

**Results for a Bi$_2$Se$_3$ nanowire without parallel magnetic field**

Let us begin by scrutinizing the case without any parallel magnetic field, $B_x = 0$. The corresponding results from the 2D surface model are shown in Fig. 2.6(a) and Fig. 2.7(a). In the single-mode regime $\mu \lesssim 78\,$meV, there is perfect T and no CAR as expected. When counterpropagating modes exist for $78\,$meV $\lesssim \mu \lesssim 102\,$meV, reflection processes (R and AR) become possible. However, as soon as the second Landau level is crossed at $\mu \approx 102\,$meV, these reflection processes are no longer possible as there are no counterpropagating modes anymore. Positive nonlocal conductance $G_{21}$ indicates that T dominates over CAR.

**Results for a Bi$_2$Se$_3$ nanowire with parallel magnetic field**

Now, we continue with the situation, where a parallel magnetic field is present: We use $B_x = 4.14\,$T such that $\phi = \phi_0/2$ and also include a vortex in the S lead, $n_v = 1$, see Section 2.2.3. Here, the single-mode regime shows perfect CAR as predicted in Section 2.4. When the counterpropagating modes are present for larger values of the chemical potential $\mu$, reflection processes are possible just as in the previous case. However, CAR still persists in this case and for all larger values of $\mu$ except that it is not the dominating transport process anymore. This can also be seen in the nonlocal conductance: $G_{21}$ is negative in the single-mode regime but becomes postive afterwards.

**Comparison with the results from the 3D BHZ model**

Next, we want to compare the results of the 2D surface model from Section 3.3.1, see Fig. 2.6(a, c) and Fig. 2.7(a), with the results of Michael Barth obtained with the 3D BHZ model [101, see also 174, 176] shown in Fig. 2.6(b, d) and Fig. 2.7(b). As one can see, the results from both models agree qualitatively very well; however, there are two quantitative differences we want to comment on: First, a slightly larger parallel magnetic field of $B_x = 4.6\,$T was used for the 3D simulations. This is due to the fact that the surface modes are not located directly on the outermost layer of sites but extend into the bulk which effectively reduces the cross section area [see appendix B of 101]. Second, the chemical potentials $\mu$ are shifted with respect to each other in the two simulations. The origin of this is that the 3D BHZ model contains an offset, i.e., a constant potential [101, 165]. Furthermore, the numerical results show a peak in the number of modes $N_1$ of lead 1 at $\mu \approx 135\,$meV. This is a numerical issue—the Landau levels are not perfectly flat in this case (note that this also happens for the 2D surface model below, see Fig. 2.8).

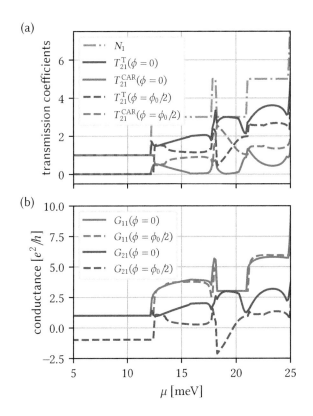

*Figure 2.8. (a) Transmission coefficients and (b) nonlocal conductance in a HgTe T junc-tion.* The parameters match the experimental values for HgTe nanowires from Ziegler et al. [133]: $\hbar v_F = 330 \, \text{meV nm}$, $w = 160 \, \text{nm}$, $h = 70 \, \text{nm}$, $\Delta = 0.25 \, \text{meV}$, and $B_z = 1 \, \text{T}$. For $\phi = \phi_0/2$, a vortex is present in the superconductor.

**Results for HgTe nanowire**

While the results above exhibit the CAR, they have one major drawback: The large magnitude of the perpendicular magnetic field $B_z$ poses a challenge for experimental realization. This can be overcome by increasing the width of the wires since, then, a smaller magnetic length $l_{B_z}$ and, thus, a smaller magnetic field $B_z$ is needed to drive the wires into the quantum Hall state. Experimentally realized HgTe nanowires are actually much wider than the dimensions from the previous section: The wires from Ziegler et al. [133] show widths from 150 nm to 500 nm. To demonstrate this, we simulate the T junction with these parameters by setting $w = l = 160$ nm, $h = 70$ nm, $\hbar v_F = 330$ meV nm, and $\Delta = 0.25$ meV. For the perpendicular magnetic field, we use $B_z = 1$ T which is enough to guarantee the quantum Hall state ($l_{B_z} = \sqrt{\hbar/eB_z} \approx 26$ nm $\ll w$; compare also the band structure of such a wire in Fig. 2.2(c)). The results are displayed in Fig. 2.8. As one can see, they show qualitative agreement with the previous ones, especially for the occurrence of T and CAR in the single-mode regime (where topological protection due to the Béri degeneracy holds).

**Outlook**

Last but not least, we want to shortly discuss some more results from Fuchs et al. [101] which have only been verified by 3D calculations of Michael Barth.

First, it is also possible that one can observe CAR and a negative nonlocal conductance over a large range of chemical potentials $\mu$ when the system is not yet in the quantum Hall state such that smaller magnitudes of $B_z$ also suffice. This is promising since it reduces the experimentally challenging requirement for large magnetic fields further.

Second, topological protection should guarantee perfect CAR in the single-mode regime in the presence of impurities and defects. This is due to the Béri degeneracy and has already been shown for the NS junction [152]. 3D simulations of Michael Barth including disorder also confirm this statement: They showed the single-mode regime to be robust against disorder and CAR to persist even outside the single-mode regime [for a more detailed discussion, we refer the reader to 101, section IV.C.].

# Chapter 3.

# Topological insulator nanowire Josephson junctions

## 3.1. Introduction

### 3D TI nanowire Josephson junctions

In this Chapter, we investigate Josephson junctions [7] made out of the 3D TI nano-wires from Chapter 2. These systems are of particular interest since the existence of (tunable) Majorana zero modes [28, 29] is predicted to give rise to a $4\pi$-periodic Josephson current [19]. Missing Shapiro steps have already been observed in experiments [30, 177] indicating the Josephson current to have a $4\pi$-periodic component. On the other hand, the geometrical setup itself is interesting and unusual: Since transport in TIs is mediated by the surface modes, these systems resemble a Josephson junction of a cylinder hosting modes with nonzero angular momentum looping around the circumference.

Here however, we focus on the critical current and its dependence on an axial magnetic field, since this is currently investigated experimentally at the University of Regensburg. In these experiments, nanowires with trapezoidal cross sections were obtained from a HgTe film by chemical etching. After etching away the appropriate parts of the capping layer, deploying Niobium (Nb) on top creates superconducting fingers on top of the wire. Figure 3.1 (a, b) shows the experimental setup of such systems, in this case for a similar experiment investigating the Shapiro spectrum [30]. Mainly two different behaviours have been observed: In some samples, the critical current declines with increasing field, whereas in other samples it oscillates with a Fraunhofer-like pattern peaking at $\phi = 0, h/4e, h/2e$, etc., see Fig. 3.1 (c). The second case is highly unusual since only a perpendicular field should give rise to a Fraunhofer pattern and, even more strikingly, no peaks should emerge when the magnetic flux $\phi$ is an odd integer mupltiple of $h/4e$, as we discuss in the next

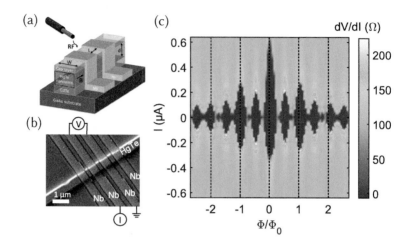

*Figure 3.1. Experiments on the 3D TI nanowire Josephson junctions:* (a) Scheme and (b) electron micrograph of the experimental setup as well as (c) the differential resistance $dV/dI$ of the sample in dependence on the current $I$ between two inner Nb stripes and the flux $\Phi$ through the nanowire cross section in units of $\Phi_0 = \phi_0/2 = h/2e$. Note that the RF antenna is not used for the current measurements. Figures (a) and (b) are taken from Fischer et al. [30][1], Fig. (c) is from R. Fischer.

paragraph. In the following, we call these peaks the $(h/4e)$-peaks.

First theoretical investigations were done by Ilan et al. [157]. They predicted the critical current to have a single maximum at $\phi = 0, h/2e, h/e, \dots$ whenever there are zero, one, two, ... vortices present along the nanowire axis. This explains the observations of the declining supercurrent. However, supercurrent oscillations with a period of $h/2e$ are also in line with their results: Any change of the magnetic field by $h/2e$ allows one vortex along the wire axis to be formed or destroyed, as described in Section 2.2.3 [see also 152], which leads to the $(h/2e)$-periodicity with $(h/2e)$-peaks. However, the $(h/4e)$-peaks remain inexplicable.

Supercurrent oscillations in dependence of an axial magnetic field have also been reported in related systems like semiconductor nanowires [178–185]. However, most of them lack the well determined periodicity of $(h/2e)$.

---

**Outline**

The objective of this work is to propose a theoretical model explaining the origin of the unusual $(h/4e)$-peaks and to identify their physical origin. This model is based on the following premises: First, there is only surface transport since we have a 3D TI. Therefore, we only model the surface states and neglect any contributions from the bulk as in Chapter 2. However, it turned out that the topological origin of the surface states is not essential—topologically trivial surface states with a quadratic dispersion also lead to $(h/4e)$-peaks. This is the reason why we consider both, topological and trivial surface states, in the following. Second, the superconducting fingers laid on top of the nanowires do not cover the hole perimeter of the nanowire cross section. Thus, we assume that the proximity induced superconductivity does not develop around the whole cross section perimeter but only in the part in contact with the superconducting fingers. This can be the case if the superconducting coherence length is smaller the the wire width; this condition is met in the systems we investigate in Sections 3.6.3 and 3.7.3. This assumption is crucial since superconductivity around the whole perimeter does not lead to $(h/4e)$-peaks as discussed above. Furthermore, the experiments show finite transparencies due to the manufacturing process, possible doping effects from the superconductor, etc. In order to include this, we include $\delta$-barriers at the NS interfaces.

In a semiclassical analysis, we dissect the different (classical) paths and calculate their contributions to the supercurrent. Among them are paths which loop around the perimeter and pick up an Aharonov-Bohm phase due to the axial magnetic field. This Aharonov-Bohm phase modifies their current phase relation leading to the $(h/4e)$-peaks whenever these paths have enough weight in comparison to the straight paths.

After introducing the geometry and Hamiltonian of our model in Section 3.2, we introduce the semiclassical method for calculating the Josephson current in Section 3.3. The current for each classical trajectory is calculated in Section 3.4. Then, we employ a minimal model to demonstrate how the $(h/4e)$-peaks emerge and get a clear picture of their physical origin. Afterwards, we turn to experimentally realizable, realistic systems with trivial and topological surface states in Section 3.6 and Section 3.7, respectively.

## 3.2. Geometry and model Hamiltonian

In this Section, we want to present the geometry of our model and its Hamiltonian. The geometry is shown in Fig. 3.2. Its base component is a nanowire or, properly speaking, the surface of a nanowire. According to experiments, we assume a rectangular cross section.[2] The junction is characterized by the following dimensions: the width $w$ and the height $h$ of the nanowire, the perimeter $P = 2(w + h)$ of the cross section, the length $C = w + 2h$ of the perimeter part covered by the superconductor,[3] the junction length $L$ and the widths $W_{L/R}$ of the left/right superconductors. All dimensions are defined in Fig. 3.2. There are two superconducting fingers laid over the nanowires such that it contacts the top and side surfaces of the nanowire but not the bottom surface. We assume the surfaces contacted by the superconductor to become superconducting themselves by the proximity effect, but not the ones which are left pristine. These regions are colored green in Fig. 3.2, whereas the noncovered parts are colored gray. Last but not least, there is a magnetic field $B$ parallel to the wire axis.

To describe the system, we employ the following Bogoliubov-de Gennes Hamiltonian:

$$H = \begin{pmatrix} h_e - \mu + U & \Delta \exp(i\varphi) \\ \Delta \exp(-i\varphi) & h_h + \mu - U \end{pmatrix}, \tag{3.1}$$

where $h_{e/h}$ are the electron and hole Hamiltonians, $\mu$ is the chemical potential, $U$ describes the barriers at the NS interfaces, $\Delta$ embodies the superconducting order parameter (and is space dependent, but real), and $\varphi$ the phase of the superconductor.

For TIs, the surface particles are captured by the Dirac Hamiltonian

$$h_{e/h} = \pm \hbar v_F \left[ \hat{k}_z \sigma_x + \left( \hat{k}_s \pm \frac{2\pi}{P} \frac{\phi}{\phi_0} \right) \sigma_y \right] \tag{3.2}$$

known from Chapter 2. To shortly recapitulate, $v_F$ is the Fermi velocity, $\hat{k}_{s,z}$ denote the wave number operators, $\sigma_{x,y,z}$ the Pauli matrices, $\phi$ refers to the magnetic flux through the cross section and $\phi_0 = h/e$ is the magnetic flux quantum. Note that we use the Hamiltonians and the magnetic field for a cylindrical nanowire; this can be done because the nanowires with circular, rectangular and trapezoidal cross sections are homeomorphic. In order to observe the ($\phi_0/4$)-periodic oszillations in the critical current, it is, however, not necessary that the surface states are of topological origin and show a Dirac dispersion (as already mentioned above). Thus, we also use the

---

[2]In experiments, the nanowires actually have a trapezoidal cross section. However, the rectangular geometry is easier to handle in theoretical models and the differences between these two geometries should be rather tiny.

[3]In Section 3.5, we allow $C$ to vary freely for illustration purposes.

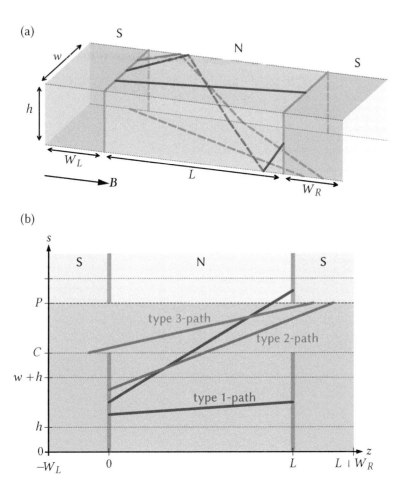

*Figure 3.2. Geometry of the 3D TI nanowire Josephson junction.* Along with a 3D sketch of the nanowire (a), we show a sketch of the unrolled 2D surface (b). The surface of the 3D TI nanowire (gray) exhibits proximity induced superconductivity (green) in two regions on the left and right; note that the wire is not superconducting around the whole perimeter but the bottom part stays normal. The barriers at certain NS interfaces are marked in orange. The different types of paths are illustrated in red, purple and blue, respectively.

quadratic Hamiltonian

$$h_{e/h} = \pm \frac{\hbar^2}{2m}\left[\hat{k}_z^2 + \left(\hat{k}_s \pm \frac{2\pi}{P}\frac{\phi}{\phi_0}\right)^2\right] \tag{3.3}$$

to simulate the junction with topologically trivial "metallic" surface states. Here, $m$ denotes the effective mass of the surface electrons. Note that the wave function $\Psi$ shares different boundary conditions in both cases: In the Dirac case, we have antiperiodic boundary conditions $\Psi(s + P) = -\Psi(s)$, whereas in the topologically trivial case, we have periodic boundary conditions $\Psi(s + P) = \Psi(s)$.

The barrier is only assumed to appear at the NS interfaces normal to the $z$ direction, see Fig. 3.2. Thus, we have

$$U(z, s) = \begin{cases} U_0[\delta(z) + \delta(z - L)] & \text{for } 0 \leq s \leq C \text{ and} \\ 0 & \text{otherwise.} \end{cases} \tag{3.4}$$

This placement is rather phenomenological to describe the imperfect transparencies observed in such systems [30]. For example, doping effects [186] would be more naturally described by increasing the chemical potential in the superconducting parts, but the introduction of a barrier shows similar effects. We also have to mention a few possible arguments underpinning the fact that the barriers are only placed on specific NS junctions: First of all, including smaller barriers at the other interfaces would show similar effects, only with a reduced overall current. Second, the side surfaces are different from the top and bottom surfaces due to the sample manufacturing; on the one hand, the additional CdTe layers on the top and bottom surfaces can lead to doping effects; on the other hand, etching away the capping layers before deploying the superconductors can also lead to differences, for example if the capping layer on top was not fully removed or if the underlaying HgTe of the top surface was attacked during the etching process.

The magnetic field parallel to the wire axis and its vectorpotential read

$$\boldsymbol{B} = B\hat{\boldsymbol{e}}_z \quad \text{and} \quad \boldsymbol{A} = \tfrac{1}{2}\boldsymbol{B}\times\boldsymbol{r} = \tfrac{1}{2}Br\hat{\boldsymbol{e}}_s. \tag{3.5}$$

The superconducting order parameter amounts to $\Delta_0$ in the regions covered by the superconducting fingers but is zero otherwise:

$$\Delta = \begin{cases} \Delta_0 & \text{for } 0 \leq s \leq C \text{ and } -W_L \leq z \leq 0, \\ \Delta_0 & \text{for } 0 \leq s \leq C \text{ and } L \leq z \leq L + W_R, \text{ and} \\ 0 & \text{otherwise.} \end{cases} \tag{3.6}$$

The phase $\varphi$ of the superconductors reads

$$\varphi = \begin{cases} +\frac{1}{2}\varphi_0 + 4\pi\frac{s}{P}\frac{\phi}{\phi_0} & \text{for } -W_L \leq z \leq 0, \\ -\frac{1}{2}\varphi_0 + 4\pi\frac{s}{P}\frac{\phi}{\phi_0} & \text{for } L \leq z \leq L + W_R, \text{ and} \\ 0 & \text{otherwise} \end{cases} \quad (3.7)$$

and consists of two parts: one governs the global phase difference $\varphi_0$ between the left and right superconductors, the other one is induced by the magnetic field; in contrast to the first one, the latter one is space dependent. In the following, we want to prove the exact form of the latter one. First, it is important to notice that the thickness of the superconducting contacts on top of the nanowires is smaller than their London penetration depth. Thus, the magnetic field penetrates the contacts and no screening supercurrents develop. This, in turn, is satisfied if the Hamiltonian is locally unitary equivalent to a Hamiltonian without a magnetic field and without a position-dependent phase. Indeed, the unitary transformation

$$U(\phi) = \exp\left( 2\pi i \frac{s}{P} \frac{\phi}{\phi_0} \tau_z \right), \quad (3.8)$$

where the Pauli matrix $\tau_z$ acts in particle-hole space, fulfills $U(\phi)H(\phi)U^\dagger(\phi) = H(0)$, see Appendix E. Note that this transformation also modifies the boundary condition for the wave function:

$$(U\Psi)(s + P) = \mp \exp\left( 2\pi i \frac{\phi}{\phi_0} \tau_z \right)(U\Psi)(s), \quad (3.9)$$

where the upper (lower) sign is for the Dirac (quadratic) Hamiltonian. This will become important in Section 3.4 when calculating the ABSs.

## 3.3. Semiclassical method

### 3.3.1. Method

For this more complex geometry, we use a semiclassical method following Ostroukh et al. [187]. This allows us to get a deeper understanding of the geometrical effect which is responsible for the $(h/4e)$-periodic supercurrent.

The semiclassical method is applicable in the limit $k_F L \gg 1$ and consists of assigning wave modes to each classical trajectory: Every classical trajectory $\Gamma$ hosts a small tube (with a width of the Fermi wavelength $\lambda_F = 2\pi/k_F$) that acts as a wave guide for one single mode. These modes are characterized by a wave number $k$ belonging to the Fermi surface (meaning $|k| = k_F$ since we have a circular Fermi surface) and contribute a current $i(\Gamma)$ to the total current.

Before showing how to obtain the total current, we want to briefly note two assumptions we make in our calculations. First, we assume straight paths neglegting any curvature. This is justified since the Lorentz force is perpendicular to the nanowire surface. Second, we will also work in the short junction limit $L \ll \xi = \hbar v_F/\Delta_0$ for simplicity. However, our findings are expected to qualitatively hold for longer junctions as well.

The total current is the integral of the current contributions $i(\Gamma)$ over all (classical) paths $\Gamma$. We choose a cut through the normal part in azimuthal direction, i.e., a straight line determined by $z = z_{\text{cut}}$ with $0 < z_{\text{cut}} < L$. Then, the integral measure is given by $ds\, dk_s/2\pi$ [187], where $s$ is the $s$-coordinate of the paths along this cut and $k_s$ is the corresponding wave number of this path. The measure can also be expressed as $ds\, d\theta \cos(\theta)k_F/2\pi$, where $\theta$ is the angle the path encloses with the $z$ direction. In equations, this reads

$$I = \frac{1}{2\pi} \int ds \int dk_s\, i(s, k_s) = \frac{k_F}{2\pi} \int ds \int d\theta\, \cos(\theta)i(s, \theta). \qquad (3.10)$$

For the rest of this Section, we explore this integral and its relationship with the geometry in more detail.

Above, we made another, bigger simplification: When barriers at the NS interfaces are present (see Fig. 3.2), they allow for normal reflection. However, normal reflection conserves the momentum parallel to the interface such that the outgoing electron does not share the same path with the incident electron. In this case, integral (3.10) is not valid anymore. Instead, one would have path integrals including all normal reflections. This is, however, very difficult to calculate since the current contribution $i$ gets harder to obtain and one would have to apply resummation techniques to solve the integrals. Since this goes beyond the scope of this work, we continue using integral (3.10) and calculate the current contribution $i$ for a 1D Josephson junction with barriers.

### 3.3.2. Classification of the paths

Looking at the geometry depicted in Fig. 3.2, one notices that we can classify the paths into different categories. Then, the integral (3.10) for the total current is evaluated seperately for each of these categories. In the rest of this Section, we want to introduce their definitions.

First, we can assign a crossing number $n$ to every path: For a given line cut $s = s_{cut}$ through the normal region, $C < s_{cut} < P$, we can count the number a path crosses this cut with a positive sign indicating a crossing in positive $s$-direction and a negative sign indicating a crossing in negative $s$-direction (it is assumed that the path runs in positive $z$ direction). In Fig. 3.2 for example, there are two red paths with $n = 0$ and $n = 1$, respectively. In fact, the crossing number can be viewed as winding number around the perimeter, i.e., the winding number of the trajectory projected onto the perimeter and closed by contracting the section covered by the superconductor to a point.

Furthermore, we can classify the paths in different types depending on which NS interface the start and end points are (see also Table 3.1 on Page 55):

**Type-1 paths** are "direct" paths. The ends of the paths are located on the $z = 0$ and $z = L$ NS interfaces. If the wire is fully superconducting, $C = P$, only these type of paths exist.

**Type-2 paths** are "mixed" paths, where one end is on a $z = const$ NS interface and the other one on a $s = const$ NS interface. We can subdivide these paths further into type-2L and type-2R paths depending on whether they start on the $z = 0$ interface or end on the $z = L$ interface (L and R denoting whether the $z = const$ interface is on the left or on the right superconductor).

**Type-3 paths** are "side" paths originating from both NS junctions at the $s = 0$ and $s = C$ interfaces.

Note that there are no paths with crossing number $n = 0$ for the type-1 and type-2 paths.

### 3.3.3. Trajectories

Without magnetic field, the dispersion of an electron in the normal part is given by

$$E = \hbar v_F \sqrt{k_z^2 + k_s^2} - \mu \tag{3.11}$$

for the Dirac Hamiltonian and by

$$E = \frac{\hbar^2}{2m}(k_z^2 + k_s^2) - \mu \tag{3.12}$$

for the quadratic Hamiltonian. Thus, the group velocity $v = \nabla_k E / \hbar$ reads

$$v = v_F \frac{k}{|k|} \tag{3.13}$$

for the Dirac and

$$v = \frac{\hbar}{2m} k \tag{3.14}$$

for the quadratic Hamiltonian; note that in both cases the direction of the velocity is given by the wave number vector $k$. Since we'll only look at paths at the Fermi surface, the wave numbers satisfy $|k| = \sqrt{k_s^2 + k_z^2} = k_F$, where $k_F$ is the Fermi wave number. Thus, the trajectory of a path can be parametrized as

$$r(t) = \left( s_0 + t \frac{k_s}{k_F}, z_0 + t \frac{k_z}{k_F} \right), \tag{3.15}$$

where we used the arc length $t$ as parameter and denoted the start point coordinates by $(s_0, z_0)$. Using the $s$ or $z$ coordinate as parameter, one gets

$$s(z) = s_0 + (z - z_0) \frac{k_s}{k_z} \quad \text{and} \quad z(s) = z_0 + (s - s_0) \frac{k_z}{k_s}, \tag{3.16}$$

respectively. If the end coordinates $(s, z) = (s_1, z_1)$ are known, the wave numbers $k_s$, $k_z$ can be determined through the relations

$$\frac{k_z}{k_s} = \frac{z_1 - z_0}{s_1 - s_0} \quad \text{and} \quad k_z^2 + k_s^2 = k_F^2. \tag{3.17}$$

Thus, the dependence of the wave numbers $k_s$ and $k_z$ on the start and end coordinates is given by

$$k_s = k_F \frac{s_1 - s_0}{\sqrt{(s_1 - s_0)^2 + (z_1 - z_0)^2}} \tag{3.18}$$

and

$$k_z = k_F \frac{z_1 - z_0}{\sqrt{(s_1 - s_0)^2 + (z_1 - z_0)^2}}, \tag{3.19}$$

respevtively. The start and end coordinates for all types of paths are given in Table 3.1.

*Table 3.1. Start and end coordinates of the paths* for different path types and crossing numbers $n$. The start coordinates $(s_0, z_0)$ and end coordinates $(s_1, z_1)$ are either constant or varied within an interval $[a, b]$.

| path type | | $s_0$ | $z_0$ | $s_1$ | $z_1$ |
|---|---|---|---|---|---|
| 1 | | $[0, C]$ | $0$ | $[nP, nP + C]$ | $L$ |
| 2L | $n > 0$ | $[0, C]$ | $0$ | $nP$ | $[L, L + W_R]$ |
| | $n < 0$ | $[0, C]$ | $0$ | $nP + C$ | $[L, L + W_R]$ |
| 2R | $n > 0$ | $C$ | $[-W_L, 0]$ | $[nP, nP + C]$ | $L$ |
| | $n < 0$ | $0$ | $[-W_L, 0]$ | $[nP, nP + C]$ | $L$ |
| 3 | $n > 0$ | $C$ | $[-W_L, 0]$ | $nP$ | $[L, L + W_R]$ |
| | $n < 0$ | $0$ | $[-W_L, 0]$ | $nP + C$ | $[L, L + W_R]$ |

### 3.3.4. Phase space integration

#### Type-1 paths

Let $I_{1,n}$ be the current of all type-1 paths with crossing number $n$. Solving the integral (3.10) by substitution with Eq. (3.18) and inserting the appropriate integral bounds, one gets

$$
\begin{aligned}
I_{1,n} &= \frac{1}{2\pi} \int_0^C ds_0 \int_{k_{s,\min}}^{k_{s,\max}} dk_s \, i_{1,n}(s_0, k_s) \\
&= \frac{1}{2\pi} \int_0^C ds_0 \int_{nP}^{nP+C} ds_1 \left( \frac{\partial}{\partial s_1} k_F \frac{s_1 - s_0}{[(s_1 - s_0)^2 + (z_1 - z_0)^2]^{1/2}} \right) i_{1,n}(s_0, s_1) \\
&= \frac{k_F}{2\pi} \int_0^C ds_0 \int_{nP}^{nP+C} ds_1 \frac{L^2}{[(s_1 - s_0)^2 + L^2]^{3/2}} i_{1,n}(s_0, s_1) \\
&= \frac{k_F}{2\pi} \int_0^C ds_0 \int_0^C ds_1 \frac{L^2}{[(nP + s_1 - s_0)^2 + L^2]^{3/2}} i_{1,n}(s_0, nP + s_1).
\end{aligned}
\tag{3.20}
$$

#### Type-2 paths

The currents of the type-2L and type-2R paths are denoted by $I_{2L,n}$ and $I_{2R,n}$, respectively. Note that the cases $n > 0$ and $n < 0$ have to be treated separately since the start and end coordinates cannot be expressed uniformly, see Table 3.1. One has

$$I'_{2L,n;n>0} = \frac{1}{2\pi} \int_0^C ds_0 \int_L^{L+W_R} dz_1 \left( \frac{\partial}{\partial z_1} k_F \frac{s_1 - s_0}{[(s_1 - s_0)^2 + (z_1 - z_0)^2]^{1/2}} \right) i_{2L,n}(s_0, z_1)$$

$$= \frac{k_F}{2\pi} \int_0^C ds_0 \int_L^{L+W_R} dz_1 \frac{(nP - s_0)z_1}{[(nP - s_0)^2 + z_1^2]^{3/2}} i_{2L,n}(s_0, z_1), \tag{3.21}$$

$$I'_{2L,n;n<0} = \frac{k_F}{2\pi} \int_0^C ds_0 \int_L^{L+W_R} dz_1 \frac{(s_0 - nP - C)z_1}{[(nP + C - s_0)^2 + z_1^2]^{3/2}} i_{2L,n}(s_0, z_1), \tag{3.22}$$

$$I'_{2R,n;n>0} = \frac{k_F}{2\pi} \int_{-W_L}^0 dz_0 \int_0^C ds_1 \frac{(nP + s_1 - C)(L - z_0)}{[(nP + s_1 - C)^2 + (L - z_0)^2]^{3/2}} i_{2R,n}(z_0, nP + s_1), \tag{3.23}$$

and

$$I'_{2R,n;n<0} = \frac{k_F}{2\pi} \int_{-W_L}^0 dz_0 \int_0^C ds_1 \frac{(-nP - s_1)(L - z_0)}{[(-nP - s_1)^2 + (L - z_0)^2]^{3/2}} i_{2R,n}(z_0, nP + s_1). \tag{3.24}$$

Note that we wrote $I'$ since this is not yet the final result: One, still, has to exclude the paths that cross any superconducting region without starting or ending there by adjusting the bounds of the integrals. For example, a type-2L path with $n > 0$ (compare the type-2 path depicted in Fig. 3.2) must not cross the right superconductor before the end point with $s_1 = nP$. This means that $s(z = L)$ has to take a value in the uncovered part below $s_1 = nP$. Thus, one has the condition $s(L) > (n - 1)P + C$. Similarly, one has the conditions $s(L) < (n + 1)P$ for type-2L paths with $n < 0$, $s(0) < P$ for type-2R paths with $n > 0$, and $s(0) > C - P$ for type-2R paths with $n < 0$. Using Eqs. (3.16) and (3.17), these conditions can be rewritten as

$$z_1 < L \frac{nP - s_0}{(n - 1)P + C - s_0} \quad \text{or} \quad s_0 > nP - (P - C) \frac{z_1}{z_1 - L} \tag{3.25}$$

for type-2L paths with $n > 0$,

$$z_1 < L \frac{nP + C - s_0}{(n + 1)P - s_0} \quad \text{or} \quad s_0 < nP + P \frac{z_1}{z_1 - L} - C \frac{L}{z_1 - L} \tag{3.26}$$

for type-2L paths with $n < 0$,

$$s_1 < P - (P - C) \frac{L}{z_0} \quad \text{or} \quad z_0 > -L \frac{P - C}{s_1 - P} \tag{3.27}$$

for type-2R paths with $n > 0$, and

$$s_1 > (P - C)\frac{L - z_0}{z_0} \qquad \text{or} \qquad z_0 > -L\frac{P - C}{-s_1 - P + C} \qquad (3.28)$$

for type-2R paths with $n < 0$. Thus, we get the following integrals:

$$I_{2L,n;n>0} = \frac{k_F}{2\pi} \int_0^C ds_0 \int_L^{b_{2L,n;n>0}(s_0)} dz_1 \frac{(nP - s_0)z_1}{[(nP - s_0)^2 + z_1^2]^{3/2}} i_{2L,n}(s_0, z_1), \qquad (3.29)$$

$$I_{2L,n;n<0} = \frac{k_F}{2\pi} \int_0^C ds_0 \int_L^{b_{2L,n;n<0}(s_0)} dz_1 \frac{(s_0 - nP - C)z_1}{[(nP + C - s_0)^2 + z_1^2]^{3/2}} i_{2L,n}(s_0, z_1), \qquad (3.30)$$

$$I_{2R,n;n>0} = \frac{k_F}{2\pi} \int_{-W_L}^0 dz_0 \int_0^{b_{2R,n;n>0}(z_0)} ds_1$$
$$\frac{(nP + s_1 - C)(L - z_0)}{[(nP + s_1 - C)^2 + (L - z_0)^2]^{3/2}} i_{2R,n}(z_0, nP + s_1), \qquad (3.31)$$

and

$$I_{2R,n;n<0} = \frac{k_F}{2\pi} \int_{-W_L}^0 dz_0 \int_{a_{2R,n;n<0}(z_0)}^C ds_1 \frac{(-nP - s_1)(L - z_0)}{[(-nP - s_1)^2 + (L - z_0)^2]^{3/2}} i_{2R,n}(z_0, nP + s_1) \qquad (3.32)$$

with the bounds

$$b_{2L,n;n>0}(s_0) = \min\left(L + W_R, L\frac{nP - s_0}{(n - 1)P + C - s_0}\right), \qquad (3.33)$$

$$b_{2L,n;n<0}(s_0) = \min\left(L + W_R, L\frac{nP + C - s_0}{(n + 1)P - s_0}\right), \qquad (3.34)$$

$$b_{2R,n;n>0}(z_0) = \min\left(C, -(n - 1)P - (P - C)\frac{L}{z_0}\right), \qquad (3.35)$$

and

$$a_{2R,n;n<0}(z_0) = \max\left(0, (P - C)\frac{L - z_0}{z_0} + nP\right). \qquad (3.36)$$

## Type-3 paths

Similar calculations lead to the current $I_{3,n}$ of the type-3 paths. However, the bounds of the integrals become slightly more intricate. For $n > 0$, one has the two conditions $s(0) < P$ and $s(L) > (n - 1)P + C$. Furthermore, these conditions can only be satisfied simultaneously for $n > 1$ if $z_0 > -L(P - C)/[(n - 2)P + C]$, such that one has three constraints on the boundaries. Similarly, one has the conditions $s(0) > C - P$ and $s(L) < (n - 1)P$ for $n < 0$ as well as $z_0 > L(P - C)/[(n + 2)P - C]$ for $n < -1$. Thus, the integrals read

$$I_{3,n} = \frac{k_F}{2\pi} \int_{a_{3,n}^{(0)}}^{0} dz_0 \int_{a_{3,n}^{(1)}(z_0)}^{b_{3,n}^{(1)}(z_0)} dz_1 \frac{(|n|P - C)^2}{[(|n|P - C)^2 + (z_1 - z_0)^2]^{3/2}} i_{3,n}(z_0, z_1) \quad (3.37)$$

with the bounds

$$a_{3,n}^{(0)} = \begin{cases} -W_L & \text{for } n = \pm 1 \text{ and} \\ \max\left(-W_L, -L \frac{P-C}{(|n|-2)P+C}\right) & \text{for } |n| > 1, \end{cases} \quad (3.38)$$

$$a_{3,n}^{(1)}(z_0) = \max\left(L, -z_0 \frac{(|n| - 1)P}{P - C}\right), \quad (3.39)$$

and

$$b_{3,n}^{(1)}(z_0) = \begin{cases} L + W_R & \text{for } n = \pm 1 \text{ and} \\ \min\left(L + W_R, L \frac{|n|P-C}{(|n|-1)P} - z_0 \frac{P-C}{(|n|-1)P}\right) & \text{for } |n| > 1. \end{cases} \quad (3.40)$$

**Comparison to a planar Josephson junction**

From the upper results, one can derive the current of a planar Josephson junction by taking the limit $C \to \infty$ of $I_{n=0}$ and using the Josephson current $i_{n=0} = (e\Delta_0/\hbar)\sin(\varphi_0/2)$ [11]. Then, $C$ corresponds to the width of the junction. Since there exist only type-1 paths for $n = 0$, one has

$$\begin{aligned} I_{n=0} = I_{1,n=0} &= \frac{k_F}{2\pi} \int_0^C ds_0 \int_0^C ds_1 \frac{L^2}{[(nP + s_1 - s_0)^2 + L^2]^{3/2}} i_{n=0} \\ &= \frac{k_F}{\pi}\left(\sqrt{L^2 + C^2} - L\right)\frac{e\Delta_0}{\hbar}\sin\left(\tfrac{1}{2}\varphi_0\right) \end{aligned} \quad (3.41)$$

which gives

$$I_{n=0} \approx k_F C \frac{e\Delta_0}{\pi\hbar}\sin\left(\tfrac{1}{2}\varphi_0\right) \quad (3.42)$$

for $C \to \infty$. This results is identical to the result from Kulik and Omel'Yanchuk [188, see also 187, 189].

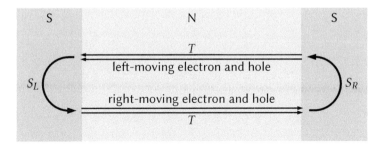

*Figure 3.3. Schematic illustration of Andreev bound states.* In an SNS junction, left moving electrons and holes in the normal part are Andreev reflected (and maybe also partially normal reflected) on the left NS interface. The reflected particles move to the right until they undergo a similar scattering process on the right NS interface ending as again left-moving particles. This allows the formation of bound states, the Andreev bound states (ABS).

## 3.4. Andreev bound states and current contributions

The current contribution of each classical trajectory can be calculated via the Andreev bound states (ABSs). In an SNS junction, ABSs embody the microscopic picture for the Josephson current [5, 8–11], see Fig. 3.3. To illustrate this, let us consider a left-moving electron in the N region with an energy inside the superconducting band gap. When it reaches the left NS interface, it is Andreev reflected as hole (if the interface is ideal) and leaves behind a Cooper pair in the left superconductor. The hole moves towards the right NS interface and is, similarly, Andreev reflected as electron, destroying a Cooper pair in the right superconductor. Interference, now, leads to the formation of bound states. During such a process, a Cooper pair is transferred from one superconductor to the other causing a supercurrent.

In this Section, we calculate the ABSs assigned to the classical trajectories including the axial magnetic flux through the wire and barriers at the left and right NS interfaces and derive the current contributions.

### 3.4.1. Andreev bound states from the scattering formalism

The scattering formalism was first utilized by Beenakker [11] to describe the Andreev bound states. In this Section, we follow Pientka et al. [190].

Transport at the left and right NS interfaces can be described by the scattering matrices $S_{L/R}$. Unitarity allows us to write

$$S_{L/R} = r_A \begin{pmatrix} rr_N & tr^*_{L/R} \\ tr_{L/R} & -rr^*_N \end{pmatrix} = \exp(i\alpha) \begin{pmatrix} r\exp(i\varphi_N) & t\exp(-i\varphi_{L/R}) \\ t\exp(i\varphi_{L/R}) & -r\exp(-i\varphi_N) \end{pmatrix}, \quad (3.43)$$

where $\alpha$ describes the Andreev reflection phase (usually, $\alpha = -\arccos(E/\Delta_0)$), $r^2$ and $t^2$ are the coefficients for normal and Andreev reflection (we have $0 \leq r, t \leq 1$ and $r^2 + t^2 = 1$), $\varphi_{L/R}$ denotes the phase of the left or right superconductor, and $\varphi_N$ describes the phase upon normal reflection. The transmission through the normal region is governed by the transmission matrix

$$T = \begin{pmatrix} t_e & 0 \\ 0 & t_h \end{pmatrix} = \begin{pmatrix} \exp(ik_e L) & 0 \\ 0 & \exp(ik_h L) \end{pmatrix}, \quad (3.44)$$

where $k_{e/h}$ represents the electron or hole wave number and $L$ is the junction length. The subgap spectrum is determined by

$$\det(1 - S_L T S_R T) = 0 \quad (3.45)$$

which is the condition for the formation of bound states, compare Fig. 3.3.

## 3.4.2. Effect of the magnetic field

To investigate the effect of the magnetic field, we use the Hamiltonian $H(\phi = 0)$ and the boundary conditions $\Psi(s + P) = \mp \exp(2\pi i(\phi/\phi_0)\tau_z)\Psi(s)$ for the wave functions, see Eqs. (3.1) and (3.9). In this form, the magnetic field only occurs in the transmission matrix $T$ and not in the scattering matrices $S_{L/R}$. Especially, the phases $\varphi_{L/R}$ of the left and right superconductors are given by $\pm\varphi_0/2$.

First, notice that the magnetic field breaks time reversal symmetry [157] such that the transmission from the left to the right and from the right to the left cannot be described with the same transmission matrix any longer. Therefore, Eq. (3.45) has to be changed to

$$\det(1 - S_L T_{LR} S_R T_{RL}) = 0 \quad (3.46)$$

with the two (distinct) transmission matrices $T_{LR}$ and $T_{RL}$.

Let's look at a path with crossing number $n$. Since the boundary condition conserves the wave number, the transmission matrices can be written as products of the transmission matrix $T$ from Eq. (3.44) and the boundary condition:

$$T_{LR} = \left[\mp \exp\left(-2\pi i \frac{\phi}{\phi_0} \tau_z\right)\right]^n T = (\mp 1)^n \begin{pmatrix} \exp(ik_e L - i\gamma) & 0 \\ 0 & \exp(ik_h L + i\gamma) \end{pmatrix} \quad (3.47)$$

and

$$T_{RL} = \left[\mp \exp\left(+2\pi i \frac{\phi}{\phi_0} \tau_z\right)\right]^n T = (\mp 1)^n \begin{pmatrix} \exp(ik_e L + i\gamma) & 0 \\ 0 & \exp(ik_h L - i\gamma) \end{pmatrix}, \quad (3.48)$$

where

$$\gamma = 2\pi n \frac{\phi}{\phi_0} \quad (3.49)$$

is exactly the Aharonov-Bohm phase

$$\gamma = \frac{e}{\hbar} \int_\Gamma \mathrm{d}\boldsymbol{l} \cdot \boldsymbol{A} \quad (3.50)$$

an electron acquires when moving from the left to the right superconductor along the trajectory $\Gamma$.

Note that the antiperiodic boundary condition belonging to the Dirac Hamiltonian (3.2) cancels out in the subgap condition (3.46) since both $T_{LR}$ and $T_{RL}$ contain the factor $(\mp 1)^n$.

Writing $f = \exp(i\gamma)$ and inserting the expressions (3.43), (3.47) and (3.48), the subgap condition (3.46) reads

$$0 = \det(1 - S_L T_{LR} S_R T_{RL}) = 1 + r_A^4 r_e^2 t_h^2 - r^2 r_A^2 [r_N^2 t_e^2 + (r_N^*)^2 t_h^2]$$
$$- t^2 r_A^2 t_e t_h [r_R r_L^* f^2 + r_R^* r_L (f^*)^2]. \quad (3.51)$$

After multiplication with $(r_A^2 t_e t_h)^*$, one gets

$$\cos(2\alpha + (k_e + k_h)L) = r^2 \cos(2\varphi_N + (k_e - k_h)L) + t^2 \cos(\varphi_L - \varphi_R - 2\gamma). \quad (3.52)$$

It is important to notice that the magnetic field manifests exclusively as modification of the phase difference $\varphi_L - \varphi_R$ with the Aharonov-Bohm phase $\gamma$ (the factor of 2 stemming from the fact that one has two particles, an electron and a counterpropagating hole, each of them contributing $\gamma$). This agrees with the expression of Ostroukh et al. [187] and with the substitution of the phase difference $\varphi_L - \varphi_R$ by its gauge invariant expression [4, p. 202].

In the case of perfect Andreev reflection, $r = 0$ and $t = 1$, and in the short-junction limit $L \to 0$, the subgap condition (3.52) is reduced to

$$\cos(2\alpha) = \cos(\varphi_L - \varphi_R - 2\gamma). \tag{3.53}$$

For $\alpha = -\arccos(E/\Delta_0)$ [11] and $\varphi_L - \varphi_R = \varphi_0$, one obtains the ABS spectrum

$$E = \pm \Delta_0 \cos\left(\tfrac{1}{2}\varphi_0 - \gamma\right) \tag{3.54}$$

in agreement with Ostroukh et al. [187].

For finite transmission, $t < 1$, we can denote the transmission coefficient with $\tau = t^2$. When the normal reflection phase vanishes, $\varphi_N = 0$, Eq. (3.52) yields the ABSs

$$E = \pm \Delta_0 \sqrt{1 - \tau \sin^2\left(\tfrac{1}{2}\varphi_0 - \gamma\right)} \tag{3.55}$$

in the short junction limit $L \to 0$. This is in agreement with [5, 11].

### 3.4.3. Asymmetric junctions

So far, we have assumed that the two NS interfaces are identical and share the same scattering matrices (up to the superconducting phase). This, however, is not true for type-2 paths. For an asymmetric junction, the scattering matrices read

$$S_{L/R} = r_{A,L/R} \begin{pmatrix} s_{L/R} r_{N,L/R} & t_{L/R} r^*_{L/R} \\ t_{L/R} r_{L/R} & -s_{L/R} r^*_{N,L/R} \end{pmatrix}$$

$$= \exp(i\alpha_{L/R}) \begin{pmatrix} s_{L/R} \exp(i\varphi_{N,L/R}) & t_{L/R} \exp(-i\varphi_{L/R}) \\ t_{L/R} \exp(i\varphi_{L/R}) & -s_{L/R} \exp(-i\varphi_{N,L/R}) \end{pmatrix} \tag{3.56}$$

with $s_{L/R}^2 + t_{L/R}^2 = 1$. Note the change of the variable name from $r$ to $s_{L/R}$ in order to avoid naming conflicts. Then, the subgap condition (3.45) becomes

$$0 = \det(1 - S_L T_{LR} S_R T_{RL}) = 1 + r_{A,L}^2 r_{A,R}^2 t_e^2 t_h^2 - r_{A,L} r_{A,R} s_L s_R [r_{N,L} r_{N,R} t_e^2 + r^*_{N,L} r^*_{N,R} t_h^2]$$
$$- r_{A,L} r_{A,R} t_L t_R [r^*_L r_R f^2 + r_L r^*_R (f^*)^2] t_e t_h. \tag{3.57}$$

After multiplication with $(r_{A,L} r_{A,R} t_e t_h)^*$, one arrives at

$$\cos(\alpha_L + \alpha_R + (k_e + k_h)L)$$
$$= s_L s_R \cos(\varphi_{N,L} + \varphi_{N,R} + (k_e - k_h)L) + t_L t_R \cos(\varphi_L - \varphi_R - 2\gamma). \tag{3.58}$$

### 3.4.4. Current

The current an ABS carries is proportional to the product of the derivative of its energy with respect to the phase difference and the distribution function [5, 11, 189, 191, a derivation can be found in 192]. Since we have two ABSs with opposite sign due to particle-hole symmetry, we have

$$i = -\frac{e}{\hbar}\frac{\partial E}{\partial \varphi_0}\tanh\left(\frac{E}{2k_BT}\right),\tag{3.59}$$

where $E$ is the energy of one of the two ABSs. Here, $k_B$ is the Boltzmann constant and $T$ the temperature. In this work, we use the zero temperature limit $T \to 0$ such that Eq. (3.59) becomes

$$i = -\frac{e}{\hbar}\frac{\partial E}{\partial \varphi_0}\operatorname{sgn}(E),\tag{3.60}$$

where sgn is the sign function,

$$\operatorname{sgn}(x) = \begin{cases} -1 & \text{for } x < 0, \\ 0 & \text{for } x = 0, \text{ and} \\ 1 & \text{for } x > 0. \end{cases}\tag{3.61}$$

Thus, the current contribution of the ABSs with $E = \pm\Delta_0\cos(\varphi_0/2-\gamma)$ from Eq. (3.54) reads

$$
\begin{aligned}
i &= -\frac{e}{\hbar}\left(\frac{\partial}{\partial\varphi_0}\Delta_0\cos\left(\tfrac{1}{2}\varphi_0 - \gamma\right)\right)\operatorname{sgn}\left(\cos\left(\tfrac{1}{2}\varphi_0 - \gamma\right)\right) \\
&= \frac{e\Delta_0}{2\hbar}\sin\left(\tfrac{1}{2}\varphi_0 - \gamma\right)\operatorname{sgn}\left(\cos\left(\tfrac{1}{2}\varphi_0 - \gamma\right)\right),
\end{aligned}\tag{3.62}
$$

whereas for finite transmission, Eq. (3.55) gives

$$
\begin{aligned}
i &= -\frac{e}{\hbar}\frac{\partial}{\partial\varphi_0}\Delta_0\sqrt{1 - \tau\sin^2\left(\tfrac{1}{2}\varphi_0 - \gamma\right)} \\
&= \frac{e\Delta_0}{4\hbar}\frac{\tau\sin(\varphi_0 - 2\gamma)}{\sqrt{1 - \tau\sin^2(\varphi_0/2 - \gamma)}}.
\end{aligned}\tag{3.63}
$$

## 3.5. Minimal model

Before discussing the more realistic setup in Sections 3.6 and 3.7, we want to look at a minimal model and investigate the dependence of the critical current on the magnetic field in this model. This will allow us to understand why and how the $(h/4e)$-peaks emerge in the experimental setup.

### 3.5.1. Model

First, we assume that there is no elastic scattering; especially, we neglect any barrier at the NS interface by setting $U_0 = 0$. We also restrict ourselves to the short junction limit $L \to 0$. Then, the ABS spectrum is given by

$$E_n = \pm\Delta_0 \cos\left(\frac{1}{2}\varphi_0 - 2\pi n\frac{\phi}{\phi_0}\right) \tag{3.64}$$

and their current contribution by

$$i_n = \frac{e\Delta_0}{2\hbar} \sin\left(\frac{1}{2}\varphi_0 - 2\pi n\frac{\phi}{\phi_0}\right) \text{sgn}\left(\cos\left(\frac{1}{2}\varphi_0 - 2\pi n\frac{\phi}{\phi_0}\right)\right). \tag{3.65}$$

Note that the ABSs and their current contribution depend on the trajectory only via the crossing number $n$, i.e., they do not depend on the detailed start and end coordinates of the trajectory.

Furthermore, we use half-infinite superconductors, $W_{L,R} \to \infty$. Together with the upper results for the ABS, this enables us to solve the integrals (3.29) to (3.32) and (3.37) analytically:

$$I_{1,n} = \frac{k_F}{2\pi}\left(\sqrt{(nP + C)^2 + L^2} - 2\sqrt{(nP)^2 + L^2} + \sqrt{(nP - C)^2 + L^2}\right)i_n, \tag{3.66}$$

$$I_{2L/R,n} = \frac{k_F}{2\pi}\left(\sqrt{(nP)^2 + L^2} - \sqrt{(|n|P - C)^2 + L^2}\right.$$
$$\left. - \sqrt{[(|n| - 1)P + C]^2 + L^2} + \sqrt{[(|n| - 1)P]^2 + L^2}\right)i_n, \tag{3.67}$$

$$I_{3,n;n=\pm1} = \frac{k_F}{2\pi}\left(\sqrt{(P - C)^2 + L^2} - L\right)i_n, \tag{3.68}$$

and

$$I_{3,n;|n|>1} = \frac{k_F}{2\pi}\left(\sqrt{(|n|P - C)^2 + L^2} - 2\sqrt{[(|n| - 1)P]^2 + L^2}\right.$$
$$\left. + \sqrt{[(|n| - 2)P + C]^2 + L^2}\right)i_n. \tag{3.69}$$

Note that the contributions of the modes get smaller the higher the crossing number $n$ is. This happens because the integral measure gets smaller the larger the angle $\theta$

with the $z$ axis is, see Eq. (3.10). Therefore, we only include trajectories with crossing numbers $n = 0, \pm 1$ in our minimal model.

With these assumptions, the total current reads

$$I = I_{1,0} + (I_{1,+1} + I_{2L,+1} + I_{2R,+1} + I_{3,+1}) + (I_{1,-1} + I_{2L,-1} + I_{2R,-1} + I_{3,-1})$$

$$= I_0 \sin\left(\frac{1}{2}\varphi_0\right) \text{sgn}\left(\cos\left(\frac{1}{2}\varphi_0\right)\right)$$

$$+ I_1 \sin\left(\frac{1}{2}\varphi_0 - 2\pi\frac{\phi}{\phi_0}\right) \text{sgn}\left(\cos\left(\frac{1}{2}\varphi_0 - 2\pi\frac{\phi}{\phi_0}\right)\right)$$

$$+ I_1 \sin\left(\frac{1}{2}\varphi_0 + 2\pi\frac{\phi}{\phi_0}\right) \text{sgn}\left(\cos\left(\frac{1}{2}\varphi_0 + 2\pi\frac{\phi}{\phi_0}\right)\right), \tag{3.70}$$

see Eq. (3.62), where $I_0$ and $I_1$ are defined as

$$I_0 = \frac{e\Delta_0}{2\hbar}\frac{k_F}{2\pi}\left(\sqrt{C^2 + L^2} - L\right) \tag{3.71}$$

and

$$I_1 = \frac{e\Delta_0}{2\hbar}\frac{k_F}{2\pi}\left(\sqrt{(P+C)^2 + L^2} - 2\sqrt{C^2 + L^2} + L\right). \tag{3.72}$$

The critical current is obtained as the maximum of $I$, $I_c = \max_{\varphi_0} I$.

## 3.5.2. Supercurrent oscillations in the minimal model

Figure 3.4 shows the current phase relation (3.70) of the minimal model for a perimeter of $P = 10L$. Since we use the zero temperature limit, there is a jump in the current phase relation whenever a sign change in any ABS occurs (which is clear from the appearance of the sgn function in Eq. (3.70)). This happens at $\varphi_0 = \pm\pi$ for the paths with crossing number $n = 0$. The paths with crossing number $n = \pm 1$, however, are shifted by the Aharonov-Bohm phase such that they show jumps at $\varphi_0 = \pm\pi + 4\pi\phi/\phi_0$ and $\varphi_0 = \pm\pi - 4\pi\phi/\phi_0$, see Fig. 3.4(a). These jumps have a direct impact on the position and value of the global maximum, compare the maxima shown as circles in Fig. 3.4(b, c). Therefore, they are responsible for the oscillations of the critical current shown in Fig. 3.5.

The jumps in the current phase relation and resulting oscillations in the critical current are similar to previous findings for semicondutor nanowires: Yokoyama, Eto, and Nazarov reported this behaviour due to the Zeeman effect [193–195] and Sriram et al. made similar observations for the orbital effect of the axial magnetic field [196].

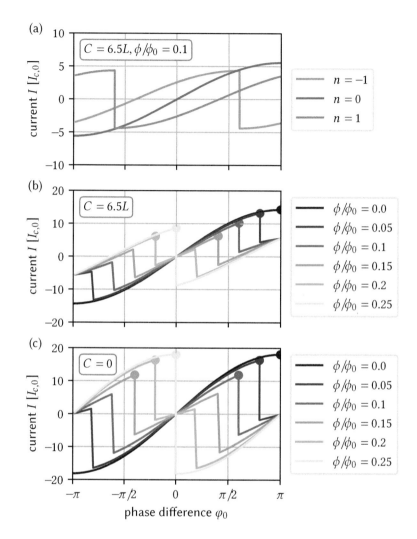

*Figure 3.4. Current phase relation of the minimal model* for a wire with perimeter $P = 10L$. The current is calculated with Eq. (3.70) and is given in units of $I_{c,0} = e\Delta_0 k_F L/4\pi\hbar$. Figure (a) shows the current contributions from the different paths with crossing number $n = 0, \pm 1$. Due to the Aharonov-Bohm phase, jumps at $\varphi_0 \neq \pm\pi$ occur for $n = \pm 1$. Figures (b) and (c) show the current phase relation for several values of the magnetic flux $\phi$ through the cross section for (b) $C = 6.5L$ and for (c) the limiting case $C = 0$, where only an infinitesimal small stripe is superconducting. The prevailing maxima are marked with circles.

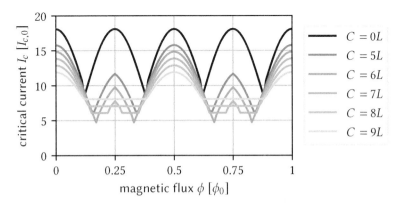

*Figure 3.5. Critical current of the minimal model* for a wire with perimeter $P = 10L$ in dependence of the magnetic flux $\phi$ through the wire cross section. The critical current is obtained as the maximum of the current phase relation (3.70) and given in units of $I_{c,0} = e\Delta_0 k_F L/4\pi\hbar$. The parameter $C$ describes the part of the cross section perimeter which is covered by the superconductor. The smaller $C$ is, the more important the paths with crossing number $n = \pm 1$ are for the total current such that additional peaks at $\phi = \phi_0/4 = h/4e$ appear.

However, in contrast to Sriram et al. [196], the peaks of the supercurrent strictly appear at multiples of $h/2e$, $h/4e$, $h/8e$ etc. in our case (depending on the maximum crossing number $|n|$ with a significant contribution to the current).

Let us now focus on the periodicity of the critical current oscillations. Since the current (3.70) is periodic in $\phi$ with a period of $\phi_0/2 = h/2e$, the critical current is also periodic in $\phi$ with the same period. To illustrate how the peaks at $\phi = h/4e$ appear, we elaborate on the unphysical but instructive limit $C \to 0$. Here, only the type-3 paths with crossing numbers $n = \pm 1$ are left and the weight $I_0$ vanishes, $I_0 = 0$. The current $I$ is $(h/2e)$-periodic in $\phi$ as above. However, it satisfies $I(\varphi_0, \phi + \phi_0/4) = I(\varphi_0 - \pi, \phi)$ since

$$
\sin\left( \frac{1}{2}\varphi_0 + 2\pi\frac{\phi}{\phi_0} + \frac{\pi}{2} \right) \mathrm{sgn}\left( \cos\left( \frac{1}{2}\varphi_0 + 2\pi\frac{\phi}{\phi_0} + \frac{\pi}{2} \right) \right)
$$

$$
= -\sin\left( \frac{1}{2}\varphi_0 + 2\pi\frac{\phi}{\phi_0} - \frac{\pi}{2} \right) \mathrm{sgn}\left( -\cos\left( \frac{1}{2}\varphi_0 + 2\pi\frac{\phi}{\phi_0} - \frac{\pi}{2} \right) \right)
$$

$$
= \sin\left( \frac{1}{2}(\varphi_0 - \pi) + 2\pi\frac{\phi}{\phi_0} \right) \mathrm{sgn}\left( \cos\left( \frac{1}{2}(\varphi_0 - \pi) + 2\pi\frac{\phi}{\phi_0} \right) \right) \quad (3.73)
$$

(this can also be seen in Fig. 3.4(c)). As the critical current is obtained as the maximum over the full interval $[-\pi, \pi]$, this results in dividing the period of the critical current in half such that it reads $h/4e$. Then, the peak at $\phi = 0$ implies a peak at $\phi = h/4e$.

For $C > 0$, these peaks remain but are less pronounced and smaller in comparison to the peaks at $\phi = 0, h/2e, \dots$ due to the coexistence of paths with $n = \pm 1$ and $n = 0$. This can be seen in Fig. 3.5. In particular, they vanish completely for $C \gtrsim 8L$ since the ratio $I_1/I_0$ becomes very small in this case: for $C = 8L$, e.g., we have $I_1/I_0 \approx 0.41$, whereas $I_1/I_0 \approx 0.96$ for $C = 6L$. Thus, the "direct paths", the type-1 paths with crossing number $n = 0$, dominate the total current more and more and the paths with crossing numbers $n = \pm 1$ become less and less important.

### 3.5.3. Implications for more realistic systems

In the last part of this Section 3.5, we want to go the other way round and address the following question: What could someone do (for example, in an experiment) such that the $(h/4e)$-peaks, $(h/8e)$-peaks, etc. emerge? For this, one has to ensure that the paths with crossing numbers $n = \pm 1$, $n = \pm 2$, etc. become more and more important. Especially, one has to ensure that the direct paths with crossing number $n = 0$ do not dominate. This can be done by, for example, adding barriers as defined in Eq. (3.4) and depicted in Fig. 3.2 which suppress their contributions. Other possibilities would be to reduce the fraction of the cross section which is covered by the superconductor, e.g., to only contact the top surface or one side surface instead of the top and both side surfaces; this amounts to reducing $C$ as we have done in Fig. 3.5. In a real experimental system, there are, of course, much more parameters one can tweak. For example, one could also try to increase the carrier density and/or mobility on the bottom surface by gating and/or doping. However, the semiclassical treatment becomes much harder if we would try to incorporate such things.

## 3.6. Semiclassical model for the quadratic Hamiltonian

### 3.6.1. Scattering matrix of the NS interface

The scattering problem of an NS junction with a quadratic Hamiltonian similar to Eq. (3.3) has been solved by Blonder, Tinkham, and Klapwijk [197]. The scattering matrix reads

$$S = \begin{pmatrix} r_{ee} & r_A \exp(i\varphi) \\ r_A \exp(-i\varphi) & r_{hh} \end{pmatrix} \tag{3.74}$$

with

$$r_A = \frac{u_0 v_0}{u_0^2 + Z^2(u_0^2 - v_0^2)} = \frac{\Delta_0}{E + i(1 + 2Z^2)(\Delta_0^2 - E^2)^{1/2}}, \tag{3.75}$$

$$r_{ee} = -\frac{(Z + i)Z(u_0^2 - v_0^2)}{u_0^2 + Z^2(u_0^2 - v_0^2)} = -\frac{2i(Z + i)Z(\Delta_0^2 - E^2)^{1/2}}{E + i(1 + 2Z^2)(\Delta_0^2 - E^2)^{1/2}}, \tag{3.76}$$

and

$$r_{hh} = -\frac{(Z - i)Z(u_0^2 - v_0^2)}{u_0^2 + Z^2(u_0^2 - v_0^2)} = -\frac{2i(Z - i)Z(\Delta_0^2 - E^2)^{1/2}}{E + i(1 + 2Z^2)(\Delta_0^2 - E^2)^{1/2}}, \tag{3.77}$$

where

$$Z(k_s) = \frac{mU_0}{\hbar^2 k_z(k_s)} = \frac{m}{\hbar^2} U_0 (k_f^2 - k_s^2)^{-1/2} \tag{3.78}$$

is the dimensionless barrier strength and

$$u_0 = \left(\frac{\Delta_0}{2E}\right)^{1/2} \exp\left(\frac{1}{2}i \arccos\left(\frac{E}{\Delta_0}\right)\right) \tag{3.79}$$

and

$$v_0 = \left(\frac{\Delta_0}{2E}\right)^{1/2} \exp\left(-\frac{1}{2}\text{i}\arccos\left(\frac{E}{\Delta_0}\right)\right) \tag{3.80}$$

are the superconducting coherence factors. Note that, as expected, $r_{ee} = 0$ and $r_{he} = \exp(-\text{i}\varphi - \text{i}\arccos(E/\Delta_0))$ when there is no barrier, $U_0 = 0$. Writing the S matrix (3.74) as in Eq. (3.43), one gets

$$S_{L/R} = \exp(\text{i}\alpha)\begin{pmatrix} r\exp(\text{i}\varphi_N) & t\exp(\mp\text{i}\varphi_0/2) \\ t\exp(\pm\text{i}\varphi_0/2) & -r\exp(-\text{i}\varphi_N) \end{pmatrix} \tag{3.81}$$

with

$$\alpha = \arg(r_A) = -\arctan\left((1 + 2Z^2)(\Delta_0^2/E^2 - 1)^{1/2}\right), \tag{3.82}$$

$$t = |r_A| = \frac{\Delta_0}{[E^2 + (1 + 2Z^2)^2(\Delta_0^2 - E^2)]^{1/2}}, \tag{3.83}$$

$$r = |r_{ee}| = (1 - t^2)^{1/2} = \frac{2(1 + Z^2)^{1/2}Z(\Delta_0^2 - E^2)^{1/2}}{[E^2 + (1 + 2Z^2)^2(\Delta_0^2 - E^2)]^{1/2}}, \tag{3.84}$$

and

$$\varphi_N = \arg(r_{ee}) - \alpha = -\arctan(Z). \tag{3.85}$$

The momenta read

$$k_{e/h} = \left(\frac{2m}{\hbar^2}(\mu \pm E) - k_s^2\right)^{1/2}. \tag{3.86}$$

### 3.6.2. Andreev bound states

The ABSs for a type-$m$ path with crossing number $n$ can be obtained by using the results from Section 3.4 with the scattering matrix from Eq. (3.74). However, we have to use appropriate values for $Z$ for the left and right scattering matrices $S_{L/R}$: type-1 paths have a barrier at both NS interfaces such that $Z$ is defined as in Eq. (3.78) for both interfaces; type-2 paths have a barrier only at one NS interface such that we have $Z$ from Eq. (3.78) for one interface and $Z = 0$ for the other; type-3 paths experience no barrier at all such that $Z = 0$ for both scattering matrices. As a result, the ABSs for a type-$m$ path with crossing number $n$ can be written as

$$E_{m,n} = \pm\Delta_0\sqrt{1 - \tau_m \sin^2\left(\frac{1}{2}\varphi_0 - 2\pi n\frac{\phi}{\phi_0}\right)}, \tag{3.87}$$

see Eqs. (3.49) and (3.55). The parameters $\tau_m$ are given by the results from Appendix F and read

$$\tau_1 = \frac{1}{1 + 4Z^2}, \quad \tau_{2L/R} = \frac{1}{1 + Z^2}, \quad \text{and} \quad \tau_3 = 1, \tag{3.88}$$

where the dimensionsless barrier strength $Z$ is defined as in Eq. (3.78).

Note that $\tau_{2L/R}$ ($\tau_1$) amounts to the transmission through a plain $\delta$-barrier of height $U_0$ ($2U_0$) without any superconductors involved. This is in perfect agreement with the thought experiment of Beenakker [11], where any barrier can be shifted into the N region by an infinitesimal amount such that one is left with a clean NS interface.

Furthermore, notice that $\tau_3 = 1$ implies that the ABSs take the form of Eq. (3.54), $E_{3,n} = \pm\Delta_0 \cos(\varphi_0/2 - 2\pi n\phi/\phi_0)$.

## 3.6.3. Supercurrent oscillations

To investigate the supercurrent oscillations, we calculate the critical current for realistic parameters: We take the width $w$ and height $h$ of the nanowire to be $w = 400\,\text{nm}$ and $h = 80\,\text{nm}$ and the width of the superconducting fingers to be $W_L = W_R = 1000\,\text{nm}$ since these values are experimentally accessible [30, 133]. Furthermore, we use $\mu = 30\,\text{meV}$ for the chemical potential and set $\hbar^2/2m = 330\,\text{meV nm}^2$ such that the energy scales are comparable with the parameters used for the Dirac Hamiltonian in Section 3.7. Then, the Fermi wave number is $k_F \approx 0.3\,\text{nm}^{-1}$ such that $k_F L \approx 3$ to $30$ for $L = 10\,\text{nm}$ to $100\,\text{nm}$ and the semiclassical limit is satisfied. For the induced order parameter $\Delta_0$, we use $\Delta_0 = 0.8\,\text{meV}$ as in Section 3.7. Then, the coherence length is $\xi \approx 250\,\text{nm}$ such that taking the short junction limit is justified.

Since the paths with larger angles are less important (they are weighted with $\cos(\theta)$ in the integral Eq. (3.10)), we introduce the cutoff $n_{\max} = 1$, only taking into account paths with crossing number $|n| \leq n_{\max} = 1$. This is a good approximation as the critical current does not change when setting $n_{\max}$ to values greater than 1, see Appendix H.

The results for the critical current are shown in Fig. 3.6 for two different lengths $L$. As expected, it is $(h/2e)$-periodic with peaks at $\phi = 0, h/2e, \ldots$ When the barrier strength $U_0$ increases, the transmissions $\tau_{1,2}$ of type-1 and type-2 paths decrease and, with them, the currents $I_{1,n}$ and $I_{2,n}$. This leads to a smaller current overall. However, the current $I_{1,0}$ gets less dominant, especially since the type-3 paths are not affected by the barrier at all such that the currents $I_{3,n}$ stay the same. This also means that the paths with crossing numbers $n = \pm 1$ get more important and peaks

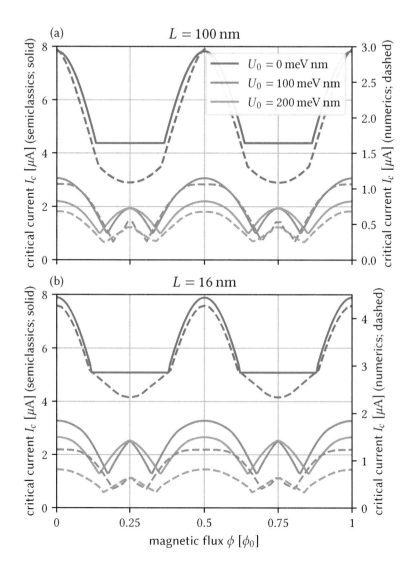

*Figure 3.6. Critical current of the nanowire Josephson junction from Fig. 3.2 with metallic surface states for two different junction lengths L. Semiclassical results are drawn with solid lines and belong to the left y axis; numerical results from Michael Barth [198] are drawn with dashed lines and belong to the right y axis. The current oscillates with a period of $\phi_0/2 = h/2e$, but shows peaks at $\phi_0/4 = h/4e$ when (sufficiently large) barriers (compare Fig. 3.2) at the NS interfaces are present.*

*Table 3.2. Quotient of the critical current maxima from semiclassics divided by the maxima from numerics with metallic surface states for different barrier strengths $U_0$ and different junction lengths $L$. See Fig. 3.6 for the full semiclassical and numerical results.*

|  | $U_0 = 0\,\text{meV nm}$ | $U_0 = 100\,\text{meV nm}$ | $U_0 = 200\,\text{meV nm}$ |
|---|---|---|---|
| for $L = 100$ nm: | 2.65 | 2.89 | 3.25 |
| for $L = 16$ nm: | 1.85 | 2.66 | 3.25 |

at $\phi = h/4e, 3h/4e, \ldots$ emerge like in Section 3.5. For very large barrier strengths, the $(h/4e)$-peaks and the $(h/2e)$-peaks are nearly equal in size meaning that the current contributions of all paths are suppressed except for the ones from the type-3 paths with $n = \pm 1$.

Figure 3.6 also includes the numerical results from Michael Barth [198]. They show the same qualitative behaviour and even share the broadening of the peaks. However, the critical currents obtained from numerics are substantially smaller; the quotients of the maxima from semiclassics and numerics are listed in Table 3.2. There are several reasons for this: First and foremost, the amplitude of the ABS spectra as obtained in the numerics is not $\Delta_0$ but, roughly estimated, $\Delta_0/2$ [198]. The reason for this is the following: at the superconducting contacts, the perimeter also forms an *SNS* junction such that ABSs form at the uncovered bottom surface. Thus, the wire at the contacts looks like there would be an "effective gap" $\Delta_{\text{eff}}$ which is smaller than the order parameter $\Delta_0$, $\Delta_{\text{eff}} < \Delta_0$. Furthermore, the issue that the semiclassical method ignores the change of direction upon normal reflection also does not exist in numerics. This agrees with the fact that the critical current decreases faster with increasing barrier strength $U_0$ in numerics than in the semiclassical calculations, compare Table 3.2. Another reason is that numerics is not bound to the short-junction limit but also captures the behaviour of ABSs for intermediate-length junctions. This explains why the quotients get smaller for shorter junction lengths $L$ like in Table 3.2. Similarly, the numerical results are for the finite temperature $T = 50\,\text{mK}$ which also slightly decreases the current.

Next, we want to discuss how the plateaus arise in the semiclassical model without barrier ($U_0 = 0$). Since $U_0 = 0$, we also have $Z = 0$ such that $\tau_m = 1$ and $E_{m,n} = \pm\Delta_0 \cos(\varphi_0/2 - 2\pi n\phi/\phi_0)$ for all types of paths ($m$ denotes the type of path). Thus, the current contributions $i_{m,n}$ are independent of the start and end points of the

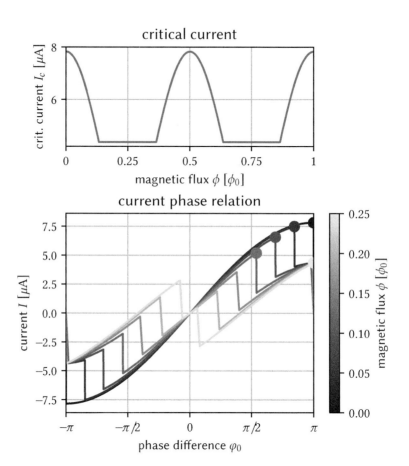

*Figure 3.7. Current phase relation of the nanowire Josephson junction from Fig. 3.2 with metallic surface states without barriers at the NS interfaces* for junction length $L =$ 100 nm. The maxima are marked with circles.

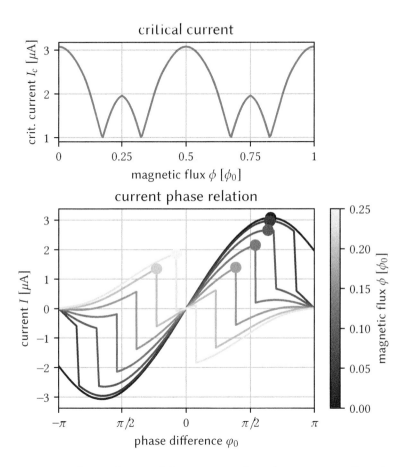

*Figure 3.8. Current phase relation of the nanowire Josephson junction from Fig. 3.2 with metallic surface states with barriers at the NS interfaces* for junction length $L = 100$ nm and barrier strength $U_0 = 100$ meV nm. The maxima are marked with circles.

trajectories such that the currents $I_{m,n}$ are proportional to

$$\sin\left(\frac{1}{2}\varphi_0 - 2\pi n\frac{\phi}{\phi_0}\right)\mathrm{sgn}\left(\cos\left(\frac{1}{2}\varphi_0 - 2\pi n\frac{\phi}{\phi_0}\right)\right). \tag{3.89}$$

Then, the current satisfies $I_{m,n}(\varphi_0) = I_{m,-n}(-\varphi_0) = -I_{m,-n}(\varphi_0)$ and one has

$$I_{m,n}(\pm\pi) + I_{m,-n}(\pm\pi) = I_{m,n}(\pm\pi) + I_{m,n}(\mp\pi) = I_{m,n}(\pm\pi) - I_{m,n}(\pm\pi) = 0 \tag{3.90}$$

for any $n > 0$ such that

$$I(\pm\pi) = I_{1,0}(\pm\pi) + I_{2,0}(\pm\pi) + I_{3,0}(\pm\pi) = I_{1,0}(\pm\pi). \tag{3.91}$$

Now, around $\phi = h/4e$ and $\phi = 3h/4e$, the maximum of the current phase relation is located at $\varphi_0 = \pi$ which is the maximum of $I_{1,0}$.[4] This can clearly be seen in the current phase relation shown in Fig. 3.7. Since $I_{1,0}$ is independent of the magnetic flux $\phi$, the critical current $I_c = I_{1,0}(\pm\pi)$ is as well leading to the plateau in the critical current. In the numerics on the other hand, the transmissions are not exactly one since the junction length $L$ is slightly bigger than 0 and the temperature is also different from zero. As a consequence, the jumps in the currents are smoothened out. Thus, the maxima are not located directly at $\varphi_0 = \pi$ and the current contributions from the paths with $n > 0$ are small but nonzero. However, they are minimal for $\phi = h/4e$ and $\phi = 3h/4e$ like the current $I(\varphi_0, \phi)$ is also minimal in the semiclassics for a fixed $\varphi_0$ slightly below $\pi$, see Fig. 3.7.

Let us now look at the current phase relation for the larger barrier strength $U_0 = 100\,\mathrm{meV\,nm}$ shown in Fig. 3.8. There, one can see that the current $I_{1,0}$ does not dominate the total current anymore but the currents $I_{n=\pm1}$ do: The maximum shifts from $\varphi_0 = \pi$ to $\varphi_0 = 0$ when $\phi$ is changed from 0 to $\phi_0/4$ following the maximum of the paths with crossing numbers $n = \pm1$ and the current changes its sign in the interval $(0, \pi)$.

## 3.7. Semiclassical model for the Dirac Hamiltonian

### 3.7.1. Scattering matrix of the NS interface

For the Dirac Hamiltonian (3.2), the scattering matrix of an NS junction can be calculated in the same way as done by [197] for the quadratic Hamiltonian. This is

---

[4]The current contribution $I_{1,0}$ dominates the total current and the maxima of the other contributions $\sum_{n\neq0} I_{m,n}(\varphi_0)$ are smaller.

done in Appendix G. The elements of the scattering matrix (3.74) are given by

$$r_A = \frac{u_0 v_0}{u_0^2 + 2Z^2(1 + Z^2)^{-1} \tan^2(\theta)(u_0^2 - v_0^2)}$$

$$= \frac{\Delta_0}{E + i[1 + 2Z^2(1 + Z^2)^{-1} \tan^2(\theta)](\Delta_0^2 - E^2)^{1/2}}, \quad (3.92)$$

$$r_{ee} = -\frac{Z[\cos(\theta) - iZ][\cos(\theta) + i\sin(\theta)]\sin(\theta)(u_0^2 - v_0^2)}{(1 + Z^2)\cos^2(\theta)u_0^2 + Z^2\sin^2(\theta)(u_0^2 - v_0^2)}$$

$$= -\frac{Z}{1 + Z^2} \frac{2i[\cos(\theta) - iZ][1 + i\tan(\theta)]\tan(\theta)(\Delta_0^2 - E^2)^{1/2}}{E + i[1 + 2Z^2(1 + Z^2)^{-1}\tan^2(\theta)](\Delta_0^2 - E^2)^{1/2}}, \quad (3.93)$$

and

$$r_{hh} = -\frac{Z[\cos(\theta) + iZ][\cos(\theta) - i\sin(\theta)]\sin(\theta)(u_0^2 - v_0^2)}{(1 + Z^2)\cos^2(\theta)u_0^2 + Z^2\sin^2(\theta)(u_0^2 - v_0^2)}$$

$$= -\frac{Z}{1 + Z^2} \frac{2i[\cos(\theta) + iZ][1 - i\tan(\theta)]\tan(\theta)(\Delta_0^2 - E^2)^{1/2}}{E + i[1 + 2Z^2(1 + Z^2)^{-1}\tan^2(\theta)](\Delta_0^2 - E^2)^{1/2}}, \quad (3.94)$$

where

$$\theta = \arctan\left(\frac{k_y}{k_z}\right) = \arctan\left(\frac{k_y}{[\mu^2(\hbar v_F)^{-2} - k_y^2]^{1/2}}\right) \quad (3.95)$$

is the angle of incidence and

$$Z = \frac{U_0}{\hbar v_F} \quad (3.96)$$

the dimensionless barrier strength.

Using the parametrization of Eqs. (3.43) and (3.81), one has

$$\alpha = \arg(r_A) = -\arctan\left([1 + 2Z^2(1 + Z^2)^{-1}\tan^2(\theta)](\Delta_0^2/E^2 - 1)^{1/2}\right), \quad (3.97)$$

$$t = |r_A| = \frac{\Delta_0}{\{E^2 + [1 + 2Z^2(1 + Z^2)^{-1}\tan^2(\theta)]^2(\Delta_0^2 - E^2)\}^{1/2}}, \quad (3.98)$$

$$r = \frac{2Z(1 + Z^2)^{-1/2}\tan(\theta)[1 + Z^2(1 + Z^2)^{-1}\tan^2(\theta)]^{1/2}(\Delta_0^2 - E^2)^{1/2}}{\{E^2 + [1 + 2Z^2(1 + Z^2)^{-1}\tan^2(\theta)]^2(\Delta_0^2 - E^2)\}^{1/2}} \quad (3.99)$$

and

$$\varphi_N = 2\arctan\left(\frac{\cos(\theta) + Z\tan(\theta)}{Z - \sin(\theta) - [1 + Z^2 + Z^2\tan^2(\theta)]^{1/2}}\right). \quad (3.100)$$

Note that $r$ can become negative in this expression. Using ordinary polar form, one would have to write

$$r = \frac{2Z(1+Z^2)^{-1/2}|\tan(\theta)|[1+Z^2(1+Z^2)^{-1}\tan^2(\theta)]^{1/2}(\Delta_0^2 - E^2)^{1/2}}{\{E^2 + [1+2Z^2(1+Z^2)^{-1}\tan^2(\theta)]^2(\Delta_0^2 - E^2)\}^{1/2}} \quad (3.101)$$

and

$$\varphi_N = \pi\Theta(-Z\tan(\theta)) + 2\arctan\left(\frac{\cos(\theta) + Z\tan(\theta)}{Z - \sin(\theta) - [1+Z^2 + Z^2\tan^2(\theta)]^{1/2}}\right). \quad (3.102)$$

### 3.7.2. Andreev bound states

The calculation of the ABS is also similar to Section 3.6.2. They can again be written in the form of Eq. (3.87), where in this case, the $\tau_m$ read

$$\tau_1 = \frac{1}{\sin^2(\varphi_N) + [1+2Z^2(1+Z^2)^{-1}\tan^2(\theta)]^2\cos^2(\varphi_N)}, \quad (3.103)$$

$$\tau_2 = \frac{1}{1 + Z^2(1+Z^2)^{-1}\tan^2(\theta)}, \quad (3.104)$$

and

$$\tau_3 = 1. \quad (3.105)$$

In contrast to the results of Section 3.6.2 for the quadratic Hamiltonian (3.3), the parameters $\tau_{1,2}$ directly depend on the angle of incidence $\theta$, whereas the dimensionless barrier strength is independent of $\theta$.

### 3.7.3. Supercurrent oscillations

For investigating the oscillations of the critical current, we look at systems with the following parameters: The width $w$ and height $h$ of the wire cross section are set to $w = 300\,\mathrm{nm}$ and $h = 80\,\mathrm{nm}$, the Junction length $L$ to $L = 200\,\mathrm{nm}$, and the widths of the superconducting fingers to $W_L = W_R = 1000\,\mathrm{nm}$. These dimensions match the experimental values [30, 133]. For the Fermi velocity, we use $\hbar v_F = 330\,\mathrm{meV\,nm}$ corresponding to HgTe nanowires [101, 133]. The chemical potential is set to $\mu = 30\,\mathrm{meV}$ which is large enough to agree with experiments (in experiments, the chemical potential is typically quite far in the conduction band) but small enough to be treatable in numerical calculations we want to compare our results to. With these values, the Fermi wave number is $k_F \approx 0.09\,\mathrm{nm^{-1}}$; since $k_F L \approx 18$, we are indeed in the semiclassical limit. The superconducting order parameter $\Delta_0$ is set to $\Delta_0 = 0.8\,\mathrm{meV}$;

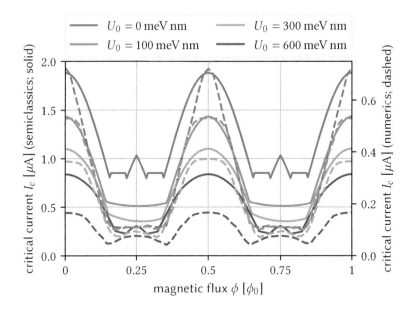

*Figure 3.9. Critical current of the 3D TI nanowire Josephson junction from Fig. 3.2 with effective Dirac surface states.* Semiclassical results are drawn with solid lines and belong to the left *y* axis; numerical results from Michael Barth [198] are drawn with dashed lines and belong to the right *y* axis. The current oscillates with a period of $\phi_0/2 = h/2e$, but shows peaks at $\phi_0/4 = h/4e$ when (sufficiently large) barriers (compare Fig. 3.2) at the NS interfaces are present.

this is in line with the experimental estimation $0.22\,\text{meV} \leq \Delta_0 \leq 0.95\,\text{meV}$ of Fischer et al. [30]. Thus, the coherence length reads $\xi \approx 400\,\text{nm}$ such that taking the short junction limit is justified.

As in the previous Section, we use $n_{\max} = 1$ and ignore all paths with crossing number $|n| > 1$; including them does not lead to big improvements, see Appendix H.

The critical current is shown in Fig. 3.9. Qualitatively, it looks very similar to the critical current obtained for the surface states with a quadratic dispersion. However, there is one difference which stands out: The $(h/4e)$-peaks also exist for $U_0 = 0$ and vanish when increasing the barrier strength initially. This behaviour is not reproduced in the numerics. Let us first discuss why we see this feature in the semiclassical results and afterwards why they deviate from the numerical results.

Figure 3.10 shows the current contributions $I_{m,n}$ for different barrier strengths $U_0$,

*Table 3.3. Quotient of the critical current maxima from semiclassics divided by the maxima from numerics with Dirac surface states for different barrier strengths $U_0$. See Fig. 3.9 for the full semiclassical and numerical results.*

| $U_0 = 0\,\text{meV nm}$ | $U_0 = 100\,\text{meV nm}$ | $U_0 = 300\,\text{meV nm}$ | $U_0 = 600\,\text{meV nm}$ |
| --- | --- | --- | --- |
| 2.60 | 2.69 | 2.94 | 5.05 |

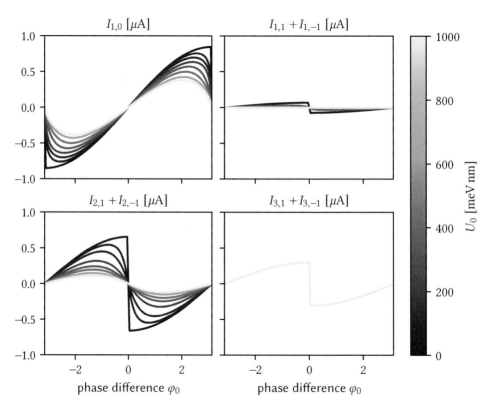

*Figure 3.10. Current contributions of the 3D TI nanowire Josephson junction with effective Dirac surface states for different barrier strengths $U_0$. The magnetic flux $\phi$ through the cross section is set to $\phi = \phi_0/4 = h/4e$. The current $I_{1,0}$ is only slightly affected by the barrier as they posses a small angle of incidence $\theta$; the currents $I_{1,\pm1}$ and $I_{2,\pm1}$ decrease strongly as expected, whereas the currents $I_{3,\pm1}$ do not change at all as the type-3 paths are not affected by the barrier.*

where the magnetic flux $\phi$ is fixed to $\phi = \phi_0/4$. When there is no barrier, $U_0 = 0$, the currents of the type-2 and type-3 paths are nearly as large as the current of the type-1 paths with $n = 0$ and, summed up, the paths with $n = \pm 1$ contribute a larger current than the type-1 paths with $n = 0$. Therefore, the critical current around $\phi = \phi_0/4$ is dominated by those paths, leading to the $(h/4e)$-peaks. In addition to that, one notices that the contributions from the type-1 paths with crossing number $n = 0$ are only slightly affected by the barrier: These paths have a very small angle of incidence $\theta$ such that the transmissions $\tau_1$ are nearly one, see Eqs. (3.103) and (3.104) (for $\theta = 0$, one has even $\tau_1 = 1$). The type-1 and type-2 paths with crossing number $n = \pm 1$ on the other hand, are stronger affected by the barrier. While their contributions are substantial for vanishing barriers ($U_0 = 0$), they become unimportant very fast such that the $(h/4e)$-peaks vanish. The contributions from the type-3 paths, however, are smaller but unaffected by the barriers such that they gain importance for very large barrier strengths $U_0$ and are responsible for the emergence of the $(h/4e)$-peaks when $U_0$ is increased again.

As already discussed in Section 3.6.3, numerics also respects the finite, nonzero junction length $L$ and the nonzero temperature. Now, type-3 and type-2 paths are typically longer than type-1 paths and paths with crossing number $n = \pm 1$ are typically longer than paths with $n = 0$. Therefore, we expect these paths to contribute less to the current than in the semiclassics. In turn, the current $I_{1,0}$ becomes more dominant preventing the $(h/4e)$-peaks for small barrier strengths $U_0$.

Last, we want to look at the current phase relation shown in Fig. 3.11. The results look smiliar to the previous results for the surface states with a quadratic dispersion (see Section 3.6.3). However, the type-1 and type-2 paths have a finite transmission $\tau_{1,2} \neq 1$ (whenever $\theta \neq 0$ and $U_0 \neq 0$). This has the consequence that there are no jumps in the currents $I_{1,n}$ and $I_{2,n}$. Therefore, the jumps at $\phi = \pm\pi$ vanish as well[5] which can be seen in Fig. 3.10.

## Concluding remarks

Our findings presented in Sections 3.5 to 3.7 suggest that the $(h/4e)$-peaks are caused by paths winding around the perimeter (i.e., the paths with crossing number $n = \pm 1$) which pick up an Aharonov-Bohm phase. Since these peaks are also present for topologically trivial surfaces states with a quadratic dispersion, see Section 3.6, we conclude that this effect is of geometrical and not of topological origin.

---

[5]For $\phi = 0, h/2e, ...$, the jumps at $\phi = \pm\pi$ still exist due to $I_{3,n}$, but are much less pronounced.

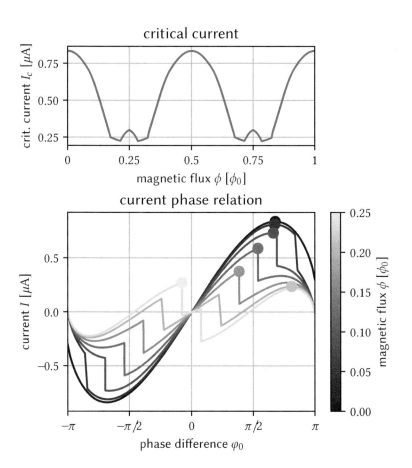

*Figure 3.11. Current phase relation of the 3D TI nanowire Josephson junction from Fig. 3.2 with effective Dirac surface states for barrier strength $U_0 = 600\,\text{meV}\,\text{nm}$. The maxima are marked with circles.*

Looking at the experimentally measured critical current in Fig. 3.1(c), one notices an overall decrease for increasing magnitute of the parallel magnetic field. This behaviour is not captured in the analytical and numerical results of Figs. 3.6 and 3.9. We attribute this to the fact that the proximity induced superconductivity is slowly destroyed with increasing magnetic field [compare 164].

# Conclusion

To conclude this thesis, let us briefly summarize what was done and review our key results.

In the first Chapter, we investigated superconducting bilayer graphene (SBLG) with a chemisorbed adatom. Such systems can exhibit subgap states called Yu-Shiba-Rusinov (YSR) states. We calculated the Green's function of SBLG, the $T$ matrix of the full system including the adatom, and, from that, the energies of the subgap states. Due to using realistic parameters for a chemisorbed hydrogen atom, we confirmed the existence of YSR states in such systems. Furthermore, we investigated the YSR state spectrum in dependence of the chemical potential $\mu$ and the exchange coupling $J$. Since YSR states have already been measured in graphene grain boundaries, our findings propose that one can detect such YSR states in SBLG in actual experiments, for example by measuring the local density of states. Furthermore, they show that it is worth to further investigate graphene systems with adatoms as discussed below.

Chapters 2 and 3 dealt with 3D topological insulator (TI) nanowires, which received a lot of experimental research in the last years. They exhibit surface states and an insulating bulk such that one can restrict oneself to a model for the surface states alone.

In Chapter 2, we proposed a T junction device—made from 3D TI nanowires, where the T arm is proximitized by an $s$-wave superconductor—as a platform for observing crossed Andreev reflection (CAR). This is interesting since the reversed process of CAR, Cooper pair splitting, can be used to generate entangled electrons. We examined the T junction working principle and established the occurence of CAR in such systems. A 2D surface model allowed the numerical simulation of its transport behavior by calculating transmission coefficients and conductances. The results contain the signatures of CAR, for example by displaying a negative nonlocal conductance. The single-mode regime renders perfect CAR over a large range of chemical potentials, but a magnetic field parallel to the T arm can be used to switch between electron transmission (T) and CAR. Comparison with (numerically more demanding) 3D calculations of Michael Barth shows agreement regarding the physical properties. This T junction device is within experimental reach (we used realistic parameters for our calculations) such that it makes up for an experimental proposal for clear-cut observation of CAR and negative nonlocal conductance.

The final Chapter deals with Josephson junctions made from 3D TI nanowires by adding $s$-wave superconductor contacts on top of them. In order to explain experimental findings, where the critical current exhibits maxima at multiples of $h/4e$, we proposed a model, where the bottom part of the wire, which is not in contact with the superconductor, is not proximitized, meaning that the superconducting pairing potential $\Delta$ is zero in these regions. The current phase relation and critical current have been calculated semiclassically by dissection into the different classical trajectories. This showed that the paths responsible for the $(h/4e)$-peaks are the ones winding once around the perimeter and, thus, picking up an Aharonov-Bohm phase. Using different Hamiltonians for the surface states revealed that the $(h/4e)$-peaks are a more universal geometrical effect of wires exhibiting surface transport with an insulating bulk—in other words, the topological origin is not necessary to observe such supercurrent oscillations in experiments. To the best of our knowledge, such geometrical phenomena have not been discussed previously in these nanowires.

Finally, let us think about possible research directions for the future.

In Chapter 1, we used Bernal stacked BLG in proximity to a superconductor. However, recent experiments reported superconductivity in magic-angle twisted bilayer graphene [199–201], twisted trilayer graphene [202, 203] and even untwisted rhombohedral trilayer graphene [204]. It is interesting to find out whether YSR states also appear in these systems, since these could be observed in experiments without the need of a superconducting substrate for introducing superconductivity. In twisted multilayer graphene, there are lots of different positions for the adatom to be chemisorbed at such that the position dependence of the YSR states promises to provide much richer physics than in untwisted multilayer graphene.

Besides, chains of magnetic adatoms on superconducting substrates have gained attraction as they might exhibit topological superconductivity and Majorana zero modes [16, 37]. Thus, it would be interesting to search for these phenomena in chains of (nonmagnetic) adatoms on SBLG or magic-angle twisted BLG.

Note that the 2D surface model used to describe the T junction of Chapter 2 can be improved further—the anisotropy of $Bi_2Se_3$, for example, has not been captured in our approach. Likewise, the model for the Josephson junction in Chapter 3 can be improved by including the finite length and finite temperature effects. However, we think that these improvements will not reveal any new physical effects. Regarding the Josephson junction, it would be much more intriguing to get a better grasp on the quantum mechanical effects of such systems, although the complex geometry of the partial superconductivity renders the analytical description very difficult. For example, the superconductors should lead to the formation of Andreev bound states (ABSs) in azimuthal direction in the uncontacted bottom part. In principle, this could phenomenologically be described by introducing an effective pairing potential $\Delta_{\text{eff}}$, but a deeper physical insight is still due. Also, the dependence of the superconducting gap on the magnetic field and the decay of the critical current for larger fields are not yet well described and can be improved.

Regarding the T junction device, experimental realization and measurement of the CAR and nonlocal conductance to confirm our theoretical findings is of great interest. Another step would be to integrate our proposed system into a larger quantum circuit to perform quantum information experiments. For example, one could operate the T junction as a Cooper pair splitter and measure the correlations between the split electrons to determine whether these electrons violate Bell (or "Bell-type") inequalities [205].

Additional experimental research on the Josephson junctions is also needed. Measurements of the current phase relation can reveal secondary maxima and, thus, provide a test for the explanation of the $(h/4e)$-peaks we proposed.

Recently, there has been substantial effort to realize Majorana fermions in solid state systems [20, 21, 206]. Although evidence of their existence has been reported in several experiments [e.g. 30, 207–211], a clear "smoking gun" experiment is still missing and the subject is under heavy debate [212–214]. As Majorana zero modes have been predicted to exist in 3D TI nanowires [28, 29, 215], these wires can provide a platform to search for further signatures of Majorana zero modes both in theory and experiment. This is necessary for additional steps in utilizing Majorana fermions for quantum computing.

# Appendices to the first chapter

## Appendix A.  Integral representations of $\tilde{g}_0$

The integral representations of all matrix elements of $\tilde{g}_0$, cf. Section 1.3.3, are given by ($z = \xi + \mu$)

$$(\tilde{g}_0)_{11} \approx \frac{V_0}{(2\pi)^2} \int_{1BZ} d^2k \, \frac{\exp(i\boldsymbol{k} \cdot (\boldsymbol{R} - \boldsymbol{R}')) \, z(z^2 - \gamma_0^2|f(\boldsymbol{k})|^2)}{(z^2 - \gamma_0^2|f(\boldsymbol{k})|^2 - z\gamma_1)(z^2 - \gamma_0^2|f(\boldsymbol{k})|^2 + z\gamma_1)}, \tag{A.1}$$

$$(\tilde{g}_0)_{12} \approx -\frac{V_0}{(2\pi)^2} \int_{1BZ} d^2k \, \frac{\exp(-i\boldsymbol{k} \cdot (\boldsymbol{R} - \boldsymbol{R}' - \boldsymbol{\delta}_1)) \, (z^2 - \gamma_0^2|f|^2)\gamma_0 f}{(z^2 - \gamma_0^2|f|^2 - z\gamma_1)(z^2 - \gamma_0^2|f|^2 + z\gamma_1)}, \tag{A.2}$$

$$(\tilde{g}_0)_{13} \approx -\frac{V_0}{(2\pi)^2} \int_{1BZ} d^2k \, \frac{\exp(-i\boldsymbol{k} \cdot (\boldsymbol{R} - \boldsymbol{R}' + \boldsymbol{\delta}_1)) \, z\gamma_1\gamma_0 f^*}{(z^2 - \gamma_0^2|f|^2 - z\gamma_1)(z^2 - \gamma_0^2|f|^2 + z\gamma_1)}, \tag{A.3}$$

$$(\tilde{g}_0)_{14} \approx \frac{V_0}{(2\pi)^2} \int_{1BZ} d^2k \, \frac{\exp(-i\boldsymbol{k} \cdot (\boldsymbol{R} - \boldsymbol{R}')) \, z^2\gamma_1}{(z^2 - \gamma_0^2|f|^2 - z\gamma_1)(z^2 - \gamma_0^2|f|^2 + z\gamma_1)}, \tag{A.4}$$

$$(\tilde{g}_0)_{21} \approx -\frac{V_0}{(2\pi)^2} \int_{1BZ} d^2k \, \frac{\exp(-i\boldsymbol{k} \cdot (\boldsymbol{R} - \boldsymbol{R}' + \boldsymbol{\delta}_1)) \, (z^2 - \gamma_0^2|f|^2)\gamma_0 f^*}{(z^2 - \gamma_0^2|f|^2 - z\gamma_1)(z^2 - \gamma_0^2|f|^2 + z\gamma_1)}, \tag{A.5}$$

$$(\tilde{g}_0)_{22} \approx \frac{V_0}{(2\pi)^2} \int_{1BZ} d^2k \, \frac{\exp(-i\boldsymbol{k} \cdot (\boldsymbol{R} - \boldsymbol{R}')) \, z(z^2 - \gamma_0^2|f|^2 - \gamma_1^2)}{(z^2 - \gamma_0^2|f|^2 - z\gamma_1)(z^2 - \gamma_0^2|f|^2 + z\gamma_1)}, \tag{A.6}$$

$$(\tilde{g}_0)_{23} \approx \frac{V_0}{(2\pi)^2} \int_{1BZ} d^2k \, \frac{\exp(\,i\boldsymbol{k} \cdot (\boldsymbol{R} - \boldsymbol{R}' + 2\boldsymbol{\delta}_1)) \, \gamma_1\gamma_0^2(f^*)^2}{(z^2 - \gamma_0^2|f|^2 - z\gamma_1)(z^2 - \gamma_0^2|f|^2 + z\gamma_1)}, \tag{A.7}$$

$$(\tilde{g}_0)_{31} \approx -\frac{V_0}{(2\pi)^2} \int_{1BZ} d^2k \, \frac{\exp(-i\boldsymbol{k} \cdot (\boldsymbol{R} - \boldsymbol{R}' - \boldsymbol{\delta}_1)) \, z\gamma_1\gamma_0 f}{(z^2 - \gamma_0^2|f|^2 - z\gamma_1)(z^2 - \gamma_0^2|f|^2 + z\gamma_1)}, \quad \text{and} \tag{A.8}$$

$$(\tilde{g}_0)_{32} \approx \frac{V_0}{(2\pi)^2} \int_{1BZ} d^2k \, \frac{\exp(-i\boldsymbol{k} \cdot (\boldsymbol{R} - \boldsymbol{R}' - 2\boldsymbol{\delta}_1)) \, \gamma_1\gamma_0^2 f^2}{(z^2 - \gamma_0^2|f|^2 - z\gamma_1)(z^2 - \gamma_0^2|f|^2 + z\gamma_1)}. \tag{A.9}$$

The other matrix elements are determined by the relations

$$(\tilde{g}_0)_{24} = (\tilde{g}_0)_{13}, \tag{A.10}$$

$$(\tilde{g}_0)_{33} = (\tilde{g}_0)_{22}, \qquad\qquad (\tilde{g}_0)_{34} = (\tilde{g}_0)_{12}, \qquad (A.11)$$

$$(\tilde{g}_0)_{41} = (\tilde{g}_0)_{14}, \qquad\qquad (\tilde{g}_0)_{42} = (\tilde{g}_0)_{31}, \qquad (A.12)$$

$$(\tilde{g}_0)_{43} = (\tilde{g}_0)_{21}, \quad \text{and} \qquad\qquad (\tilde{g}_0)_{44} = (\tilde{g}_0)_{11}. \qquad (A.13)$$

# Appendix B. Some useful mathematical identities

## Approximation of $f$

Since $\pm K \cdot \delta_1 = 0$, $\pm K \cdot \delta_2 = \mp 2\pi/3$, and $\pm K \cdot \delta_3 = \pm 2\pi/3$, the function $f$ can be approximated by

$$
\begin{aligned}
f(\pm K + q) &= \exp\left(-\mathrm{i}\frac{a|q|}{\sqrt{3}}\cos\left(\varphi_q - \frac{\pi}{2}\right)\right) \\
&\quad + \exp\left(\pm\mathrm{i}\frac{2\pi}{3}\right)\exp\left(-\mathrm{i}\frac{a|q|}{\sqrt{3}}\cos\left(\varphi_q - \frac{\pi}{2} - \frac{2\pi}{3}\right)\right) \\
&\quad + \exp\left(\mp\mathrm{i}\frac{2\pi}{3}\right)\exp\left(-\mathrm{i}\frac{a|q|}{\sqrt{3}}\cos\left(\varphi_q - \frac{\pi}{2} - \frac{4\pi}{3}\right)\right) \\
&= 1 - \mathrm{i}\frac{a|q|}{\sqrt{3}}\cos\left(\varphi_q - \frac{\pi}{2}\right) + O((a|q|)^2) \\
&\quad + \left[-\frac{1}{2} \pm \mathrm{i}\frac{\sqrt{3}}{2}\right]\left[1 + \mathrm{i}\frac{a|q|}{2\sqrt{3}}\cos\left(\varphi_q - \frac{\pi}{2}\right) - \mathrm{i}\frac{a|q|}{2}\sin\left(\varphi_q - \frac{\pi}{2}\right) + O((a|q|)^2)\right] \\
&\quad + \left[-\frac{1}{2} \mp \mathrm{i}\frac{\sqrt{3}}{2}\right]\left[1 + \mathrm{i}\frac{a|q|}{2\sqrt{3}}\cos\left(\varphi_q - \frac{\pi}{2}\right) + \mathrm{i}\frac{a|q|}{2}\sin\left(\varphi_q - \frac{\pi}{2}\right) + O((a|q|)^2)\right] \\
&= \mp\frac{3a|q|}{2\sqrt{3}}\exp(\pm\mathrm{i}\varphi_q) + O((a|q|)^2).
\end{aligned}
\tag{B.1}
$$

## Partial fraction decompositions

One has the following partial fraction decompositions:

$$
\frac{1}{(\tilde{z}_+^2 - q^2)(\tilde{z}_-^2 - q^2)} = \frac{1}{\tilde{z}_+^2 - \tilde{z}_-^2}\left(\frac{1}{\tilde{z}_-^2 - q^2} - \frac{1}{\tilde{z}_+^2 - q^2}\right)
\tag{B.2}
$$

and

$$
\frac{q^2}{(\tilde{z}_+^2 - q^2)(\tilde{z}_-^2 - q^2)} = \frac{1}{\tilde{z}_+^2 - \tilde{z}_-^2}\left(\frac{\tilde{z}_-^2}{\tilde{z}_-^2 - q^2} - \frac{\tilde{z}_+^2}{\tilde{z}_+^2 - q^2}\right).
\tag{B.3}
$$

## Calculation of the integrals

Since one has to be careful with the branch cut, we split the integrand into real and imaginary part. With $z = a + ib$, $a, b \in \mathbf{R}$,

$$
\begin{aligned}
I_1(z, q) &= \int dq \, \frac{q}{z^2 - q^2} \\
&= \frac{1}{2} \int dq \, \left( \frac{a - ib - q}{(a - q)^2 + b^2} - \frac{a - ib + q}{(a + q)^2 + b^2} \right) \\
&= -\frac{1}{4} \ln\left( (a - q)^2 + b^2 \right) + \frac{1}{2} i \arctan\left( \frac{a - q}{b} \right) \\
&\quad - \frac{1}{4} \ln\left( (a + q)^2 + b^2 \right) + \frac{1}{2} i \arctan\left( \frac{a + q}{b} \right) \\
&= -\frac{1}{4} \ln\left( [\mathrm{Re}(z^2) - q^2]^2 + [\mathrm{Im}(z^2)]^2 \right) \\
&\quad + \frac{1}{2} i \arctan\left( \frac{a - q}{b} \right) + \frac{1}{2} i \arctan\left( \frac{a + q}{b} \right) \quad (B.4)
\end{aligned}
$$

since

$$
\frac{\partial}{\partial q} \ln\left( (a \pm q)^2 + b^2 \right) = \pm \frac{2(a \pm q)}{(a \pm q)^2 + b^2} \quad (B.5)
$$

and

$$
\frac{\partial}{\partial q} \arctan\left( \frac{a \pm q}{b} \right) = \pm \frac{b}{(a \pm q)^2 + b^2}. \quad (B.6)
$$

Similarly,

$$
I_3(z, q) = \int dq \, \frac{q^3}{z^2 - q^2} = \int dq \left( q - \frac{z^2 q}{z^2 - q^2} \right) = -\frac{1}{2} q^2 + z^2 I_1(z, q). \quad (B.7)
$$

# Appendices to the second chapter

## Appendix C. Numerical implementation and finite difference method

### Finite difference method

As mentioned above, the finite difference method gives a possibility to convert the low-energy continuous models to tight-binding models. Let us demonstrate this method in one dimension. On a lattice with lattice spacing $a$ (where $x_i = x_0 + ia$), one can approximate the derivatives by the following finite differences:

$$\partial_x \psi(x_i) \approx \frac{1}{2a}(\psi(x_{i+1}) - \psi(x_{i-1})) \tag{C.1}$$

and

$$\partial_x^2 \psi(x_i) \approx \frac{1}{a^2}(\psi(x_{i+1}) - 2\psi(x_i) + \psi(x_{i-1})). \tag{C.2}$$

Thus, the operators $\hat{k}_x$ and $\hat{k}_x^2$ translate to

$$\hat{k}_x \rightarrow -\frac{i}{2a} \sum_i (|x_{i+1}\rangle\langle x_i| - |x_i\rangle\langle x_{i+1}|) \tag{C.3}$$

and

$$\hat{k}_x^2 \rightarrow \frac{1}{a^2} \sum_i (-2|x_i\rangle\langle x_i| + |x_{i+1}\rangle\langle x_i| + |x_i\rangle\langle x_{i+1}|). \tag{C.4}$$

### Fermion doubling

Note that this method has one major drawback: Due to the Nielsen-Ninomiya theorem [216–218], a second Dirac cone appears at the edge of the Brillouin zone—a phenomenon known as fermion doubling [see, e.g., 161, 219]. There exist several proposals how to deal with this [e.g. 219–225]. The most simple of them is the following [161]: Use an even number of lattice points and divide the resulting transport coefficients by a factor of two (or by a factor of four for 2D systems). This works as long as there is no scattering between the two Dirac cones. Another method consists of adding an additional mass term to the Hamiltonian [219–221]:

$$H_{\hat{n}} = \hbar v_F(\sigma \times \hat{k}) \cdot \hat{n} - \mu - \frac{1}{4}E_{bc}a^2\left(\hat{k}^2 - (\hat{k} \cdot \hat{n})^2\right)(\sigma \cdot \hat{n}) \tag{C.5}$$

so that, for example,

$$H_{\hat{z}} = \hbar v_F(\hat{k}_y\sigma_x - \hat{k}_x\sigma_y) - \mu - \frac{1}{4}E_{bc}a^2(\hat{k}_x^2 + \hat{k}_y^2)\sigma_z. \tag{C.6}$$

This introduces a gap of $E_{bc}$ to the spurious Dirac cones.

**Peierls substitution**

In tight-binding simulations, the minimal coupling $\hat{k} \rightarrow \hat{k} + eA/\hbar$ fails. Instead, one has to apply the Peierls substitution [226] to include magnetic fields. Then, any hopping $t_{i,j}$ from $x_i$ to $x_j$ is multiplied by a phase factor related to the vector potential $A$:

$$t_{i,j} \rightarrow t_{i,j} \exp\left(-i\frac{e}{\hbar}\int_{x_i}^{x_j} dl \cdot A\right). \tag{C.7}$$

**Geometrical layout of the tight-binding system**

When defining the tight-binding system in the numerical calculations, it is best to roll out the two nanowire surfaces and use a square grid for both of them. The periodic boundary conditions connect the edges of the slabs from the rolled out nanowire; hoppings between the two grids glue the two nanowires together.

**Matching conditions**

Let us now look at one edge of a rolled out nanowire to investigate how the matching conditions modify the finite differences. Let $s$ be the coordinate perpendicular to this edge and $s_k$ its position. Then, the matching condition reads $\psi_{\hat{n}_1}(s_{k+1}) = U_{\hat{n}_1\hat{n}_2}\psi_{\hat{n}_2}(s_{k+1})$. The finite difference at the edge gives

$$\partial_x\psi_{\hat{n}_1}(s_k) \approx \frac{1}{2a}\left(\psi_{\hat{n}_1}(s_{k+1}) - \psi_{\hat{n}_1}(s_{k-1})\right) = \frac{1}{2a}\left(U_{\hat{n}_2\hat{n}_1}^{\dagger}\psi_{\hat{n}_2}(s_{k+1}) - \psi_{\hat{n}_1}(s_{k-1})\right) \tag{C.8}$$

which means that the edge hoppings have to be multliplied by $U_{\hat{n}_1\hat{n}_2}$ and $U_{\hat{n}_1\hat{n}_2}^{\dagger}$, respectively: When $t_{k,k-1} = -(i/2a)\hbar v_F\sigma_n$, the edge hoppings read

$$t_{k+1,k} = -\frac{i}{2a}\hbar v_F\sigma_n U_{\hat{n}_2\hat{n}_1}^{\dagger} \quad \text{and} \quad t_{k,k+1} = \frac{i}{2a}\hbar v_F U_{\hat{n}_2\hat{n}_1}\sigma_n. \tag{C.9}$$

# Appendices to the third chapter

## Appendix D.  Alternative derivation of the
## superconducting phase

In this Appendix, we want to show an alternative derivation of the magnetic field-dependence of the superconducting phase $\varphi$ from Eq. (3.7) using the Ginzburg-Landau theory.

First, note that the order parameter is proportional to the Ginzburg-Landau wave-function $\psi$ [4, p. 111]. Neglegting the contributions of the normal state, the free energy density $f$ reads

$$f = \alpha|\psi|^2 + \frac{1}{2}\beta|\psi|^4 + \frac{1}{2m}|(-i\hbar\nabla - 2e\mathbf{A})\psi|^2. \tag{D.1}$$

Separating the absolute value and phase of the wavefunction, one can write $\psi = \psi_0 \exp(i\varphi)$ where $\psi_0 = |\psi|$, such that

$$f = \alpha\psi_0^2 + \frac{1}{2}\beta\psi_0^4 + \frac{1}{2m}\left[\hbar^2(\nabla\psi_0)^2 + (\hbar\nabla\varphi - 2e\mathbf{A})^2\psi_0^2\right]. \tag{D.2}$$

Assuming a constant order parameter, $\psi_0 = const$, the free energy becomes minimal when the term $(\hbar\nabla\varphi - 2e\mathbf{A})$ is zero [see also 227, 228]. Note that this term is proportional to the electric current density

$$\mathbf{j}_S = \frac{2e}{m} \mathrm{Re}(\psi^*(-i\hbar\nabla - 2e\mathbf{A})\psi) = -2\frac{en_S}{m}(\hbar\nabla\varphi - 2e\mathbf{A}) \tag{D.3}$$

where $n_S$ is the superconducting density such that our argumentation in Section 3.2 that there is no supercurrent is valid. In our case, this means

$$\nabla\varphi = \frac{2e}{\hbar}\mathbf{A} = \frac{e}{\hbar}Br\mathbf{e}_s \tag{D.4}$$

which is satisfied by

$$\varphi = \frac{e}{\hbar}Br^2\frac{2\pi s}{P} = 4\pi\frac{\phi}{\phi_0}\frac{s}{P} \tag{D.5}$$

for a cylindrical nanowire.

## Appendix E.  Unitary transformation of the Hamiltonian

For any operators $a$ and $b$, one has

$$[a, b^n] = \sum_{k<n} b^{n-k-1}[a,b]b^k. \tag{E.1}$$

This implies that

$$[a, \exp(\beta b)] = \sum_n \frac{1}{n!}[a, (\beta b)^n] = \beta[a,b]\exp(\beta b) \tag{E.2}$$

for any complex number $\beta$ whenever $[a,b]$ and $b$ commute with each other.

To prove that the unitary transformation (3.8) indeed fulfills $U(\phi)H(\phi)U^\dagger(\phi) = H(0)$, it is sufficient to show that

$$\exp\left(\pm 2\pi i \frac{s}{P}\frac{\phi}{\phi_0}\right) h_{e/h}(\phi) \exp\left(\mp 2\pi i \frac{s}{P}\frac{\phi}{\phi_0}\right) = h_{e/h}(0). \tag{E.3}$$

This, in turn, holds if

$$\exp\left(\pm 2\pi i \frac{s}{P}\frac{\phi}{\phi_0}\right)\left(\hat{k}_s \pm \frac{2\pi}{P}\frac{\phi}{\phi_0}\right)\exp\left(\mp 2\pi i \frac{s}{P}\frac{\phi}{\phi_0}\right) = \hat{k}_s. \tag{E.4}$$

Since $\hat{k}_s$ and $s$ do not commute with each other but satisfy $[s, \hat{k}_s] = i$ due to the canonical commutation relations, we can use Eq. (E.2) to obtain

$$\begin{aligned}
\exp&\left(\pm 2\pi i \frac{s}{P}\frac{\phi}{\phi_0}\right)\left(\hat{k}_s \pm \frac{2\pi}{P}\frac{\phi}{\phi_0}\right)\exp\left(\mp 2\pi i \frac{s}{P}\frac{\phi}{\phi_0}\right) \\
&= \left(\hat{k}_s \pm \frac{2\pi}{P}\frac{\phi}{\phi_0}\right) + \exp\left(\pm 2\pi i \frac{s}{P}\frac{\phi}{\phi_0}\right)\left[\hat{k}_s, \exp\left(\mp 2\pi i \frac{s}{P}\frac{\phi}{\phi_0}\right)\right] \\
&= \left(\hat{k}_s \pm \frac{2\pi}{P}\frac{\phi}{\phi_0}\right) + \exp\left(\pm 2\pi i \frac{s}{P}\frac{\phi}{\phi_0}\right)\left(\mp 2\pi \frac{1}{P}\frac{\phi}{\phi_0}\right)\exp\left(\mp 2\pi i \frac{s}{P}\frac{\phi}{\phi_0}\right) \\
&= \hat{k}_s
\end{aligned} \tag{E.5}$$

which completes our proof. Finally, the boundary condition (3.9) follows from

$$U(s+P)\Psi(s+P) = \mp U(P)U(s)\Psi(s) = \mp \exp\left(2\pi i \frac{\phi}{\phi_0}\tau_z\right)U(s)\Psi(s) \tag{E.6}$$

where the upper (lower) sign is for the Dirac (quadratic) Hamiltonian.

# Appendix F. Andreev bound states for imperfect interfaces

Here, we derive the form of the ABS when one or both NS interfaces of the Josephson junction are imperfect. We assume that the Andreev reflection phase and the reflection amplitude are energy dependent via

$$\alpha = -\arctan(A(\Delta_0^2/E^2 - 1)^{1/2}) \quad \text{and} \quad r^2 = \frac{\Delta_0^2}{E^2 + A^2(\Delta_0^2 - E^2)} \tag{F.1}$$

and that the normal reflection phase does not depend on the energy. This corresponds to the scattering matrices from Sections 3.6 and 3.7. Furthermore, we restrict ourselves to the short junction limit $L \to 0$.

## F.1. Two imperfect interfaces

When there are two imperfect interfaces (sharing the same scattering matrix), the subgap equation reads

$$\cos(2\alpha) = (1 - r^2)\cos(2\varphi_N) + r^2\cos(\varphi_0) \tag{F.2}$$

as derived in Section 3.4, see Eq. (3.52). After inserting $\alpha$ and $r^2$ from Eq. (F.1) and multiplying with $E^2 + A^2(\Delta_0^2 - E^2)$, one arrives at

$$E^2 - A^2(\Delta_0^2 - E^2) = [E^2 + A^2(\Delta_0^2 - E^2) - \Delta_0^2]\cos(2\varphi_N) + \Delta_0^2\cos(\varphi_0). \tag{F.3}$$

Solving this equation for $E^2$ leads to the ABSs

$$E = \pm\Delta_0\sqrt{1 - \frac{\sin^2(\varphi_0/2)}{\sin^2(\varphi_N) + A^2\cos^2(\varphi_N)}} \tag{F.4}$$

which corresponds to the results of Eq. (3.55) with $\tau = \sin^2(\varphi_N) + A^2\cos^2(\varphi_N)$.

## F.2. One imperfect interface

When there is one imperfect and one clean interface, the clean one is described by $\alpha' = -\arccos(E/\Delta_0)$ and $r = 0$. The subgap equation following from Eq. (3.58) reads

$$\cos(\alpha + \alpha') = \cos(\alpha)\cos(\alpha') - \sin(\alpha)\sin(\alpha') = r\cos(\varphi_0). \tag{F.5}$$

Restricting to $E > 0$ (which is allowed since the scattering matrix is also derived for this case), one can derive the equation

$$E^2 - A(\Delta_0^2 - E^2) = \Delta_0^2 \cos(\varphi_0). \tag{F.6}$$

Thus, the ABS read

$$E = \pm\Delta_0 \sqrt{1 - \frac{2\sin^2(\varphi_0/2)}{1 + A}} \tag{F.7}$$

which, again, corresponds to Eq. (3.55) with $\tau = (1 + A)/2$.

# Appendix G.  Scattering matrix of a TI NS junction

In this Appendix, we derive the scattering matrix from Section 3.7.1 of an NS interface with a Dirac Hamiltonian. Such a system can be described by the Hamiltonian

$$H = \begin{pmatrix} h_e & \Delta \\ \Delta^* & -h_e^* \end{pmatrix}, \quad h_e = \hbar v_F(k_x \sigma_x + k_y \sigma_y) - \mu + U_0\delta(x), \tag{G.1}$$

where

$$\Delta = \Delta_0 \exp(i\varphi)\Theta(x), \tag{G.2}$$

such that the superconductor covers the $x > 0$ half-plane. We restrict ourselves to the subgap regime $|E| \leq \Delta_0$ and to the case of positive energies $E > 0$.

## G.1.  Scattering states

The incoming and outgoing scattering states are denoted by $\Psi_{in}$ and $\Psi_{out}$, respectively, and read

$$\Psi_{in} = a_e(2\pi\hbar v_e)^{-1/2} \exp(ik_e x + ik_y y) \begin{pmatrix} \chi_{e,r} \\ 0 \end{pmatrix}$$

$$+ a_h(2\pi\hbar v_h)^{-1/2} \exp(-ik_h x + ik_y y) \begin{pmatrix} 0 \\ \chi_{h,r} \end{pmatrix} \tag{G.3}$$

and

$$\Psi_{out} = b_e(2\pi\hbar v_e)^{-1/2} \exp(-ik_e x + ik_y y) \begin{pmatrix} \chi_{e,l} \\ 0 \end{pmatrix}$$

$$+ b_h(2\pi\hbar v_h)^{-1/2} \exp(ik_h x + ik_y y) \begin{pmatrix} 0 \\ \chi_{h,l} \end{pmatrix}. \tag{G.4}$$

Their momenta are given by

$$k_e = [(E + \mu)^2(\hbar v_F)^{-2} - k_s^2]^{1/2} \quad \text{and} \quad k_h = [(E - \mu)^2(\hbar v_F)^{-2} - k_s^2]^{1/2}, \quad \text{(G.5)}$$

their group velocities by

$$v_e = v_F[1 - \hbar^2 v_F^2 k_y^2(E + \mu)^{-2}]^{1/2} \quad \text{and} \quad v_h = v_F[1 - \hbar^2 v_F^2 k_y^2(E - \mu)^{-2}]^{1/2}, \quad \text{(G.6)}$$

and their spinors are

$$\chi_{e,r/l} = \frac{1}{\sqrt{2}} \begin{pmatrix} 1 \\ (\pm k_e + ik_y)(k_e^2 + k_y^2)^{-1/2} \end{pmatrix} \quad \text{(G.7)}$$

and

$$\chi_{h,r/l} = \frac{1}{\sqrt{2}} \begin{pmatrix} 1 \\ (\mp k_h + ik_y)(k_h^2 + k_y^2)^{-1/2} \end{pmatrix}. \quad \text{(G.8)}$$

The scattering states $\Psi_{ev}$ of the evanescent particles in the superconductor are

$$\Psi_{ev} = c_{el}(2\pi\hbar v_F)^{-1/2} \exp(ik_{el}x + ik_y y) \begin{pmatrix} f u_0 \chi_{el,r} \\ f^* v_0 \chi_{el,r} \end{pmatrix}$$

$$+ c_{hl}(2\pi\hbar v_F)^{-1/2} \exp(-ik_{hl}x + ik_y y) \begin{pmatrix} f v_0 \chi_{hl,r} \\ f^* u_0 \chi_{hl,r} \end{pmatrix} \quad \text{(G.9)}$$

and have the wave numbers

$$k_{el} = [(\mu + i\xi)^2(\hbar v_F)^{-2} - k_s^2]^{1/2} \quad \text{and} \quad k_{hl} = [(\mu - i\xi)^2(\hbar v_F)^{-2} - k_s^2]^{1/2} \quad \text{(G.10)}$$

where

$$\xi = (\Delta_0^2 - E^2)^{1/2} \quad \text{and} \quad f = \exp(i\varphi/2). \quad \text{(G.11)}$$

The coherence factors read

$$u_0 = \left(\frac{\Delta_0}{2E}\right)^{1/2} \exp\left(\frac{1}{2}i \arccos\left(\frac{E}{\Delta_0}\right)\right) \quad \text{(G.12)}$$

and

$$v_0 = \left(\frac{\Delta_0}{2E}\right)^{1/2} \exp\left(-\frac{1}{2}i \arccos\left(\frac{E}{\Delta_0}\right)\right) \quad \text{(G.13)}$$

and the spinors

$$\chi_{el,r/l} = \frac{1}{\sqrt{2}} \begin{pmatrix} 1 \\ (\pm k_{el} + ik_y)(k_{el}^2 + k_y^2)^{-1/2} \end{pmatrix} \tag{G.14}$$

and

$$\chi_{hl,r/l} = \frac{1}{\sqrt{2}} \begin{pmatrix} 1 \\ (\mp k_{hl} + ik_y)(k_{hl}^2 + k_y^2)^{-1/2} \end{pmatrix}. \tag{G.15}$$

Note that we did not norm the evanescent scattering states (G.9) with the actual group velocities $v_{el/hl}$ but instead with the Fermi velocity $v_F$ since their transmission coefficients are not important in the subgap regime $|E| < \Delta_0$.

## G.2. Matching condition

To determine the matching conditions at $x = 0$, one has to integrate the eigenequation of the Hamiltonian. This leads to

$$\lim_{\substack{x \to 0, \\ x>0}} \begin{pmatrix} -i\hbar v_F \sigma_x & 0 \\ 0 & i\hbar v_F \sigma_x \end{pmatrix} \Psi(x)$$

$$= \lim_{\substack{x \to 0, \\ x<0}} \begin{pmatrix} -i\hbar v_F \sigma_x - U_0 & 0 \\ 0 & i\hbar v_F \sigma_x + U_0 \end{pmatrix} \Psi(x) \tag{G.16}$$

where we assumed that the barrier only acts in the normal part (when we use a symmetric barrier, we get the exact same result for the scattering matrix). After indroducing the dimensionless barrier strength

$$Z = \frac{U_0}{\hbar v_F}, \tag{G.17}$$

the matching condition (G.16) can be transformed to

$$\Psi_{in}(0) + \Psi_{out}(0) = (X - iY\sigma_x)\Psi_{ev}(0) \tag{G.18}$$

with

$$X = \frac{1}{1 + Z^2} \quad \text{and} \quad Y = \frac{Z}{1 + Z^2}. \tag{G.19}$$

## G.3. Andreev approximation

As done by Blonder, Tinkham, and Klapwijk for the quadratic Hamiltonian [197], we utilize the Andreev approximation which is valid for $E, \Delta_0 \ll \mu$. In this approximation, we can substitute the electron and hole momenta with the Fermi momentum $k_f$ in $x$ direction,

$$k_{e/h} \approx k_{el/hl} \approx k_f = [\mu^2 (\hbar v_F)^{-2} - k_y^2]^{1/2}, \tag{G.20}$$

and the velocity with the Fermi velocity $v_f$ in $x$ direction,

$$v_{e/h} \approx v_f = v_F [1 - \hbar^2 v_F^2 k_y^2 \mu^{-2}]^{1/2}. \tag{G.21}$$

This also means that one can approximate the spinors by

$$\chi_{e,r/l} \approx \chi_{h,l/r} \approx \chi_{el,r/l} \approx \chi_{hl,l/r} \approx \chi_{f\pm} = \frac{1}{\sqrt{2}} \begin{pmatrix} 1 \\ (\pm k_f + ik_y)(k_f^2 + k_y^2)^{-1/2} \end{pmatrix}. \tag{G.22}$$

In total, the scattering states read in the Andreev reflection

$$\begin{aligned} \Psi_{in} &= a_e (2\pi\hbar v_f)^{-1/2} \exp(ik_f x + ik_y y) \begin{pmatrix} \chi_{f+} \\ 0 \end{pmatrix} \\ &\quad + a_h (2\pi\hbar v_f)^{-1/2} \exp(-ik_f x + ik_y y) \begin{pmatrix} 0 \\ \chi_{f-} \end{pmatrix}, \end{aligned} \tag{G.23}$$

$$\begin{aligned} \Psi_{out} &= b_e (2\pi\hbar v_f)^{-1/2} \exp(ik_f x + ik_y y) \begin{pmatrix} \chi_{f-} \\ 0 \end{pmatrix} \\ &\quad + b_h (2\pi\hbar v_f)^{-1/2} \exp(-ik_f x + ik_y y) \begin{pmatrix} 0 \\ \chi_{f+} \end{pmatrix}, \end{aligned} \tag{G.24}$$

and

$$\begin{aligned} \Psi_{trans} &= c_{el} (2\pi\hbar v)^{-1/2} \exp(ik_f x + ik_y y) \begin{pmatrix} f u_0 \chi_{f+} \\ f^* v_0 \chi_{f+} \end{pmatrix} \\ &\quad + c_{hl} (2\pi\hbar v)^{-1/2} \exp(-ik_f x + ik_y y) \begin{pmatrix} f v_0 \chi_{f-} \\ f^* u_0 \chi_{f-} \end{pmatrix}. \end{aligned} \tag{G.25}$$

## G.4. Scattering matrix

Inserting the scattering states Eqs. (G.3), (G.4) and (G.9) into the matching condition (G.18), one obtains the equation system

$$a_e(2\pi\hbar v_e)^{-1/2}\chi_{e,r} + b_e(2\pi\hbar v_e)^{-1/2}\chi_{e,l}$$

$$= c_{el}(2\pi\hbar v_F)^{-1/2}fu_0(X + iY\sigma_x)\chi_{el,r}$$

$$+ \frac{1}{1 + Z^2}c_{hl}(2\pi\hbar v_F)^{-1/2}fv_0(X + iY\sigma_x)\chi_{hl,r} \quad \text{(G.26)}$$

$$a_h(2\pi\hbar v_h)^{-1/2}\chi_{h,r} + b_h(2\pi\hbar v_h)^{-1/2}\chi_{h,l}$$

$$= c_{el}(2\pi\hbar v_F)^{-1/2}f^*v_0(X + iY\sigma_x)\chi_{el,r}$$

$$+ \frac{1}{1 + Z^2}c_{hl}(2\pi\hbar v_F)^{-1/2}f^*u_0(X + iY\sigma_x)\chi_{hl,r}. \quad \text{(G.27)}$$

Here, we can apply the Andreev approximation. To simplify notation, we define

$$\zeta_\pm = \frac{k_x \pm ik_y}{(k_x^2 + k_y^2)^{1/2}} = \exp(\pm i\theta) \quad \text{and} \quad \theta = \arctan\left(\frac{k_y}{k_x}\right) \quad \text{(G.28)}$$

and introduce the scaled coefficients

$$\tilde{c}_{el/hl} = (1 - \hbar^2 v^2 k_y^2 \mu^{-2})^{1/4} c_{el/hl}. \quad \text{(G.29)}$$

Thus, one arrives at the following equation system:

$$a_e + b_e = (X + iY\zeta_{f+})fu_0\tilde{c}_{el} + (X - iY\zeta_{f-})fv_0\tilde{c}_{hl} \quad \text{(G.30)}$$

$$\zeta_{f+}a_e - \zeta_{f-}b_e = (X\zeta_{f+} + iY)fu_0\tilde{c}_{el} + (-X\zeta_{f-} + iY)fv_0\tilde{c}_{hl} \quad \text{(G.31)}$$

$$a_h + b_h = (X + iY\zeta_{f+})f^*v_0\tilde{c}_{el} + (X - iY\zeta_{f-})f^*u_0\tilde{c}_{hl} \quad \text{(G.32)}$$

$$-\zeta_{f-}a_h + \zeta_{f+}b_h = (X\zeta_{f+} + iY)f^*v_0\tilde{c}_{el} + (-X\zeta_{f-} + iY)f^*u_0\tilde{c}_{hl}. \quad \text{(G.33)}$$

In order to obtain the scattering matrix, we write it in the matrix form

$$Mc = Ha \quad \text{(G.34)}$$

with the coefficient vectors

$$a = \begin{pmatrix} a_e \\ a_h \end{pmatrix} \quad \text{and} \quad c = \begin{pmatrix} b_e \\ b_h \\ \tilde{c}_{el} \\ \tilde{c}_{hl} \end{pmatrix} \quad \text{(G.35)}$$

and the matrices

$$
M = \begin{vmatrix}
-1 & 0 & (X + iY\zeta_{f+})f u_0 & (X - iY\zeta_{f-})f v_0 \\
0 & -1 & (X + iY\zeta_{f+})f^* v_0 & (X - iY\zeta_{f-})f^* u_0 \\
\zeta_{f-} & 0 & (X\zeta_{f+} + iY)f u_0 & (-X\zeta_{f-} + iY)f v_0 \\
0 & -\zeta_{f+} & (X\zeta_{f+} + iY)f^* v_0 & (-X\zeta_{f-} + iY)f^* u_0
\end{vmatrix}
\tag{G.36}
$$

and

$$
H = \begin{vmatrix}
1 & 0 \\
0 & 1 \\
\zeta_{f+} & 0 \\
0 & -\zeta_{f-}
\end{vmatrix}.
\tag{G.37}
$$

From this, one can easily calculate the (modified extended) scattering matrix $\tilde{S}$ as

$$
\tilde{S} = M^{-1}H = \begin{pmatrix} A & B \\ C & D \end{pmatrix}^{-1} \begin{pmatrix} 1 \\ G \end{pmatrix} = \begin{pmatrix} BF^{-1}(C+G) - 1 \\ F^{-1}(C+G) \end{pmatrix}
\tag{G.38}
$$

where $A, B, C, D$ and $G$ are the block matrices of $M$ and $H$ and

$$
F = D - CA^{-1}B = D + CB
\tag{G.39}
$$

is the Schur complement of $M$. Note that $S = BF^{-1}(C+G) - 1$ is the scattering matrix we need to calculate the ABS. At the end, one gets

$$
S_{11} = \frac{4Y[X\cos(\theta) - iY][\cos(\theta) + i\sin(\theta)]\sin(\theta)(u_0^2 - v_0^2)}{\det(F)},
\tag{G.40}
$$

$$
S_{12} = -\frac{4(X^2 + Y^2)\cos^2(\theta)f^2 u_0 v_0}{\det(F)},
\tag{G.41}
$$

$$
S_{21} = -\frac{4(X^2 + Y^2)\cos^2(\theta)(f^*)^2 u_0 v_0}{\det(F)},
\tag{G.42}
$$

and

$$
S_{22} = \frac{4Y[X\cos(\theta) + iY][\cos(\theta) - i\sin(\theta)]\sin(\theta)(u_0^2 - v_0^2)}{\det(F)}
\tag{G.43}
$$

with

$$
\det(F) = -4(X^2 + Y^2)\cos^2(\theta)u_0^2 - 4Y^2\sin^2(\theta)(u_0^2 - v_0^2).
\tag{G.44}
$$

# Appendix H. Critical current in dependence of the maximum crossing number

In this Appendix, we want to justify the neglection of the paths with crossing number $|n| > n_{max} = 1$ when calculating the critical current in Sections 3.6.3 and 3.7.3. For this, we repeat the calculations of the critical current using different values of $n_{max}$.

The results for the quadratic Hamiltonian, where we chose the length $L = 100$ nm, are shown in Fig. H.1. As one can see, there is no visible difference for $n_{max} = 1$ and $n_{max} = 2$. Since the weights in the integrals (3.20), (3.29) to (3.32) and (3.37) get smaller the higher $|n|$ is, we conclude that even higher values of $n$ also will not change the result and $n_{max} = 1$ is fully sufficient.

For the Dirac Hamiltonian, the critical current for different values of $n_{max}$ is shown in Fig. H.2. Here, there is no visible difference for $n_{max} > 2$ but slight differences between $n_{max} = 1$ and $n_{max} = 2$ meaning that the current contributions from the paths with crossing number $n = \pm 2$ are still noticeable. First, the current around the minima without barrier ($U_0 = 0$) is smoother and does not show dips. Thus, we conclude that the results are not fully converged for $n_{max} = 1$. Second, the current gets enlarged around $\phi = 0, h/2e, \ldots$. This has the following reason: At the above-mentioned values, the maxima for all crossing numbers $n$ align such that the current is enlarged. However, the maxima of the contributions from the paths with crossing number $n = \pm 2$ shift very fast when the magnetic flux $\phi$ moves away from these values such that the current enhancement also vanishes rapidly. Lastly, the peaks at $\phi = h/4e$ and $\phi = 3h/4e$ are already slightly visible for smaller barrier strengths $U_0$. This is due to the fact that the critical current of the paths with $n = \pm 2$ also shows these ($h/4e$)-peaks and, thus, favours the formation of these. However, all in all these effects are quite small such that setting $n_{max} = 1$ is not too much of a simplification and captures (at least) the qualitative behaviour very well.

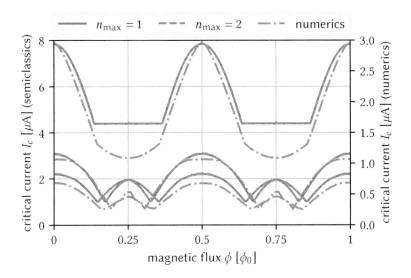

*Figure H.1. Critical current of the nanowire Josephson junction from Fig. 3.2 with metallic surface states for different maximum numbers of crossing $n_{max}$. From top to bottom, the barrier strengths are $U_0 = 0\,$meV nm, $U_0 = 100\,$meV nm, and $U_0 = 200\,$meV nm like in Fig. 3.6. Including paths with two crossings or more does not alter the results.*

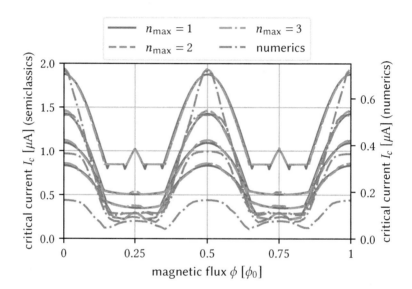

*Figure H.2. Critical current of the 3D TI nanowire Josephson junction from Fig. 3.2 with effective Dirac surface states for different maximum numbers of crossing $n_{max}$. From top to bottom, the barrier strengths are $U_0 = 0\,\text{meV}\,\text{nm}$, $U_0 = 100\,\text{meV}\,\text{nm}$, $U_0 = 300\,\text{meV}\,\text{nm}$, and $U_0 = 600\,\text{meV}\,\text{nm}$ like in Fig. 3.9. Including paths with two crossings smoothens the curve for $U_0 = 0\,\text{meV}\,\text{nm}$ and enlarges the currents at $\phi = 0, \phi_0/4, \phi_0/2, ...$; including paths with three crossings or more does not change the results.*

# Bibliography

[1] A. Castro Neto, F. Guinea, and N. M. Peres, "Drawing conclusions from graphene", Physics World **19**, 33 (2006).

[2] M. I. Katsnelson, "Graphene: carbon in two dimensions", Materials Today **10**, 20 (2007).

[3] M. Z. Hasan and C. L. Kane, "Colloquium: Topological insulators", Reviews of Modern Physics **82**, 3045 (2010).

[4] M. Tinkham, *Introduction to superconductivity*, 2nd ed. (McGraw-Hill, Inc., 1996).

[5] T. M. Klapwijk, "Proximity Effect From an Andreev Perspective", Journal of Superconductivity **17**, 593 (2004).

[6] A. Andreev, "Thermal Conductivity of the Intermediate State in Superconductors", Zhurnal Eksperimental'noi i Teoreticheskoi Fiziki **46**, 1823 (1964); [Soviet Physics JETP **19**, 1228 (1964)].

[7] B. Josephson, "Possible new effects in superconductive tunnelling", Physics Letters **1**, 251 (1962).

[8] I. O. Kulik, "Macroscopic Quantization and the Proximity Effect in S-N-S Junctions", Zhurnal Eksperimental'noi i Teoreticheskoi Fiziki **57**, 1745 (1970); [Soviet Physics JETP **30**, 944 (1970)].

[9] C. Ishii, "Josephson Currents through Junctions with Normal Metal Barriers", Progress of Theoretical Physics **44**, 1525 (1970).

[10] J. Bardeen, R. Kümmel, A. E. Jacobs, and L. Tewordt, "Structure of Vortex Lines in Pure Superconductors", Physical Review **187**, 556 (1969).

[11] C. W. J. Beenakker, "Universal limit of critical-current fluctuations in mesoscopic Josephson junctions", Physical Review Letters **67**, 3836 (1991); erratum Physical Review Letters **68**, 1442 (1992).

[12] L. Yu, "Bound State in Superconductors with Paramagnetic Impurities", Acta Physica Sinica **21**, 75 (1965).

[13] H. Shiba, "Classical Spins in Superconductors", Progress of Theoretical Physics **40**, 435 (1968).

[14] A. I. Rusinov, "Superconductivity near a paramagnetic impurity", Pis'ma v Zhurnal Eksperimental'noi i Teoreticheskoi Fiziki **9**, 146 (1968); [JETP Letters **9**, 85 (1969)].

[15] A. I. Rusinov, "On the Theory of Gapless Superconductivity in Alloys Containing Paramagnetic Impurities", Zhurnal Eksperimental'noi i Teoreticheskoi Fiziki **56**, 2047 (1969); [Soviet Physics JETP **29**, 1101 (1969)].

[16] B. W. Heinrich, J. I. Pascual, and K. J. Franke, "Single magnetic adsorbates on s -wave superconductors", Progress in Surface Science **93**, 1 (2018).

[17] X.-L. Qi and S.-C. Zhang, "Topological insulators and superconductors", Rev. Mod. Phys. **83**, 1057 (2011).

[18] M. Sato and Y. Ando, "Topological superconductors: a review", Reports on Progress in Physics **80**, 076501 (2017).

[19] A. Y. Kitaev, "Unpaired Majorana fermions in quantum wires", Physics-Uspekhi **44**, 131 (2001).

[20] J. Alicea, "New directions in the pursuit of Majorana fermions in solid state systems", Reports on Progress in Physics **75**, 076501 (2012).

[21] C. Beenakker, "Search for Majorana Fermions in Superconductors", Annual Review of Condensed Matter Physics **4**, 113 (2013).

[22] E. Cortés-del Río, J. L. Lado, V. Cherkez, P. Mallet, J.-Y. Veuillen, J. C. Cuevas, J. M. Gómez-Rodríguez, J. Fernández-Rossier, and I. Brihuega, "Observation of Yu–Shiba–Rusinov States in Superconducting Graphene", Advanced Materials **33**, 2008113 (2021).

[23] D. Kochan, M. Barth, A. Costa, K. Richter, and J. Fabian, "Spin Relaxation in s-Wave Superconductors in the Presence of Resonant Spin-Flip Scatterers", Physical Review Letters **125**, 087001 (2020).

[24] M. Barth, J. Fuchs, and D. Kochan, "Spin relaxation, Josephson effect, and Yu-Shiba-Rusinov states in superconducting bilayer graphene", Physical Review B **105**, 205409 (2022).

[25] G. B. Lesovik, T. Martin, and G. Blatter, "Electronic entanglement in the vicinity of a superconductor", The European Physical Journal B - Condensed Matter and Complex Systems **24**, 287 (2001).

[26] L. Hofstetter, S. Csonka, J. Nygård, and C. Schönenberger, "Cooper pair splitter realized in a two-quantum-dot Y-junction", Nature **461**, 960 (2009).

[27] L. G. Herrmann, F. Portier, P. Roche, A. L. Yeyati, T. Kontos, and C. Strunk, "Carbon Nanotubes as Cooper-Pair Beam Splitters", Physical Review Letters **104**, 026801 (2010).

[28] A. Cook and M. Franz, "Majorana fermions in a topological-insulator nanowire proximity-coupled to an s-wave superconductor", Physical Review B **84**, 201105 (2011).

[29] A. M. Cook, M. M. Vazifeh, and M. Franz, "Stability of Majorana fermions in proximity-coupled topological insulator nanowires", Physical Review B **86**, 155431 (2012).

[30] R. Fischer, J. Picó-Cortés, W. Himmler, G. Platero, M. Grifoni, D. A. Kozlov, N. N. Mikhailov, S. A. Dvoretsky, C. Strunk, and D. Weiss, "$4\pi$-periodic supercurrent tuned by an axial magnetic flux in topological insulator nanowires", Physical Review Research **4**, 013087 (2022).

[31] N. P. Outreach, *Press release, The Nobel Prize in Physics 1973*, (Oct. 23, 1973) https://www.nobelprize.org/prizes/physics/1973/press-release/ (visited on 05/22/2022).

[32] A. V. Balatsky, I. Vekhter, and J.-X. Zhu, "Impurity-induced states in conventional and unconventional superconductors", Reviews of Modern Physics **78**, 373 (2006).

[33] A. Yazdani, B. A. Jones, C. P. Lutz, M. F. Crommie, and D. M. Eigler, "Probing the Local Effects of Magnetic Impurities on Superconductivity", Science **275**, 1767 (1997).

[34] S.-H. Ji, T. Zhang, Y.-S. Fu, X. Chen, X.-C. Ma, J. Li, W.-H. Duan, J.-F. Jia, and Q.-K. Xue, "High-Resolution Scanning Tunneling Spectroscopy of Magnetic Impurity Induced Bound States in the Superconducting Gap of Pb Thin Films", Physical Review Letters **100**, 226801 (2008).

[35] S.-H. Ji, T. Zhang, Y.-S. Fu, X. Chen, J.-F. Jia, Q.-K. Xue, and X.-C. Ma, "Application of magnetic atom induced bound states in superconducting gap for chemical identification of single magnetic atoms", Applied Physics Letters **96**, 073113 (2010).

[36] K. J. Franke, G. Schulze, and J. I. Pascual, "Competition of Superconducting Phenomena and Kondo Screening at the Nanoscale", Science **332**, 940 (2011).

[37] S. Nadj-Perge, I. K. Drozdov, J. Li, H. Chen, S. Jeon, J. Seo, A. H. MacDonald, B. A. Bernevig, and A. Yazdani, "Observation of Majorana fermions in ferromagnetic atomic chains on a superconductor", Science **346**, 602 (2014).

[38] M. Ruby, F. Pientka, Y. Peng, F. von Oppen, B. W. Heinrich, and K. J. Franke, "End States and Subgap Structure in Proximity-Coupled Chains of Magnetic Adatoms", Physical Review Letters **115**, 197204 (2015).

[39] H. Kim, A. Palacio-Morales, T. Posske, L. Rózsa, K. Palotás, L. Szunyogh, M. Thorwart, and R. Wiesendanger, "Toward tailoring Majorana bound states in artificially constructed magnetic atom chains on elemental superconductors", Science Advances **4**, eaar5251 (2018).

[40] J. F. Steiner, C. Mora, K. J. Franke, and F. von Oppen, "Quantum Magnetism and Topological Superconductivity in Yu-Shiba-Rusinov Chains", Physical Review Letters **128**, 036801 (2022).

[41] E. Liebhaber, L. M. Rütten, G. Reecht, J. F. Steiner, S. Rohlf, K. Rossnagel, F. von Oppen, and K. J. Franke, "Quantum spins and hybridization in artificially-constructed chains of magnetic adatoms on a superconductor", Nature Communications **13**, 2160 (2022).

[42] A. Palacio-Morales, E. Mascot, S. Cocklin, H. Kim, S. Rachel, D. K. Morr, and R. Wiesendanger, "Atomic-scale interface engineering of Majorana edge modes in a 2D magnet-superconductor hybrid system", Science Advances **5**, eaav6600 (2019).

[43] S. Kezilebieke, M. N. Huda, V. Vaňo, M. Aapro, S. C. Ganguli, O. J. Silveira, S. Głodzik, A. S. Foster, T. Ojanen, and P. Liljeroth, "Topological superconductivity in a van der Waals heterostructure", Nature **588**, 424 (2020).

[44] J. O. Island, R. Gaudenzi, J. de Bruijckere, E. Burzurí, C. Franco, M. Mas-Torrent, C. Rovira, J. Veciana, T. M. Klapwijk, R. Aguado, and H. S. J. van der Zant, "Proximity-Induced Shiba States in a Molecular Junction", Physical Review Letters **118**, 117001 (2017).

[45] S. Das Sarma, M. Freedman, and C. Nayak, "Majorana zero modes and topological quantum computation", npj Quantum Information **1**, 15001 (2015).

[46] D. Chatzopoulos, D. Cho, K. M. Bastiaans, G. O. Steffensen, D. Bouwmeester, A. Akbari, G. Gu, J. Paaske, B. M. Andersen, and M. P. Allan, "Spatially dispersing Yu-Shiba-Rusinov states in the unconventional superconductor $FeTe_{0.55}Se_{0.45}$", Nature Communications **12**, 298 (2021).

[47]  P. Fan, F. Yang, G. Qian, H. Chen, Y.-Y. Zhang, G. Li, Z. Huang, Y. Xing, L. Kong, W. Liu, K. Jiang, C. Shen, S. Du, J. Schneeloch, R. Zhong, G. Gu, Z. Wang, H. Ding, and H.-J. Gao, "Observation of magnetic adatom-induced Majorana vortex and its hybridization with field-induced Majorana vortex in an iron-based superconductor", Nature Communications **12**, 1348 (2021).

[48]  D. Wang, J. Wiebe, R. Zhong, G. Gu, and R. Wiesendanger, "Spin-Polarized Yu-Shiba-Rusinov States in an Iron-Based Superconductor", Physical Review Letters **126**, 076802 (2021).

[49]  A. K. Geim and K. S. Novoselov, "The rise of graphene", Nature Materials **6**, 183 (2007).

[50]  V. P. Gusynin, S. G. Sharapov, and J. P. Carbotte, "AC conductivity of graphene: from tight-binding model to 2 + 1-dimensional quantum electrodynamics", International Journal of Modern Physics B **21**, 4611 (2007).

[51]  M. Katsnelson and K. Novoselov, "Graphene: New bridge between condensed matter physics and quantum electrodynamics", Solid State Communications **143**, Exploring graphene, 3 (2007).

[52]  C. W. J. Beenakker, "Colloquium: Andreev reflection and Klein tunneling in graphene", Reviews of Modern Physics **80**, 1337 (2008).

[53]  A. H. Castro Neto, F. Guinea, N. M. R. Peres, K. S. Novoselov, and A. K. Geim, "The electronic properties of graphene", Review of Modern Physics **81**, 109 (2009).

[54]  P. R. Wallace, "The Band Theory of Graphite", Physical Review **71**, 622 (1947).

[55]  K. S. Novoselov, A. K. Geim, S. V. Morozov, D. Jiang, Y. Zhang, S. V. Dubonos, I. V. Grigorieva, and A. A. Firsov, "Electric Field Effect in Atomically Thin Carbon Films", Science **306**, 666 (2004).

[56]  "October 22, 2004: Discovery of Graphene", APS News **18**, edited by A. Chodos (2009).

[57]  K. S. Novoselov, D. Jiang, F. Schedin, T. J. Booth, V. V. Khotkevich, S. V. Morozov, and A. K. Geim, "Two-dimensional atomic crystals", Proceedings of the National Academy of Sciences of the United States of America **102**, 10451 (2005).

[58]  N. P. Outreach, *Press release, Graphene – the perfect atomic lattice*, (Oct. 5, 2010) https://www.nobelprize.org/prizes/physics/2010/press-release/ (visited on 03/17/2022).

[59] The Royal Swedish Academy of Sciences, *Graphene, Scientific Background on the Nobel Prize in Physics 2010*, (Oct. 5, 2010) `https://www.nobelprize.org/uploads/2018/06/advanced-physicsprize2010.pdf` (visited on 03/18/2022).

[60] The Royal Swedish Academy of Sciences, *Graphene – the perfect atomic lattice*, (Oct. 5, 2010) `https://www.nobelprize.org/uploads/2018/06/popular-physicsprize2010.pdf` (visited on 03/18/2022).

[61] E. V. Castro, K. S. Novoselov, S. V. Morozov, N. M. R. Peres, J. M. B. L. dos Santos, J. Nilsson, F. Guinea, A. K. Geim, and A. H. C. Neto, "Biased Bilayer Graphene: Semiconductor with a Gap Tunable by the Electric Field Effect", Physical Review Letters **99**, 216802 (2007).

[62] A. B. Kuzmenko, E. van Heumen, D. van der Marel, P. Lerch, P. Blake, K. S. Novoselov, and A. K. Geim, "Infrared spectroscopy of electronic bands in bilayer graphene", Physical Review B **79**, 115441 (2009).

[63] S. V. Morozov, K. S. Novoselov, M. I. Katsnelson, F. Schedin, D. C. Elias, J. A. Jaszczak, and A. K. Geim, "Giant Intrinsic Carrier Mobilities in Graphene and Its Bilayer", Physical Review Letters **100**, 016602 (2008).

[64] K. S. Novoselov, E. McCann, S. V. Morozov, V. I. Fal'ko, M. I. Katsnelson, U. Zeitler, D. Jiang, F. Schedin, and A. K. Geim, "Unconventional quantum Hall effect and Berry's phase of $2\pi$ in bilayer graphene", Nature Physics **2**, 177 (2006).

[65] T. Ohta, A. Bostwick, T. Seyller, K. Horn, and E. Rotenberg, "Controlling the Electronic Structure of Bilayer Graphene", Science **313**, 951 (2006).

[66] W. Han, R. K. Kawakami, M. Gmitra, and J. Fabian, "Graphene spintronics", Nature Nanotechnology **9**, 794 (2014).

[67] S. Roche, J. Åkerman, B. Beschoten, J.-C. Charlier, M. Chshiev, S. Prasad Dash, B. Dlubak, J. Fabian, A. Fert, M. Guimarães, F. Guinea, I. Grigorieva, C. Schönenberger, P. Seneor, C. Stampfer, S. O. Valenzuela, X. Waintal, and B. van Wees, "Graphene spintronics: the European Flagship perspective", 2D Materials **2**, 030202 (2015).

[68] A. Avsar, H. Ochoa, F. Guinea, B. Özyilmaz, B. J. van Wees, and I. J. Vera-Marun, "Colloquium: Spintronics in graphene and other two-dimensional materials", Reviews of Modern Physics **92**, 021003 (2020).

[69] J. F. Sierra, J. Fabian, R. K. Kawakami, S. Roche, and S. O. Valenzuela, "Van der Waals heterostructures for spintronics and opto-spintronics", Nature Nanotechnology, 10.1038/s41565-021-00936-x (2021).

[70] I. Žutić, A. Matos-Abiague, B. Scharf, H. Dery, and K. Belashchenko, "Proximitized materials", Materials Today **22**, 85 (2019).

[71] J. Ingla-Aynés, M. H. D. Guimarães, R. J. Meijerink, P. J. Zomer, and B. J. van Wees, "24-$\mu$ spin relaxation length in boron nitride encapsulated bilayer graphene", Physical Review B **92**, 201410 (2015).

[72] M. Eschrig, "Spin-polarized supercurrents for spintronics", Physics Today **64**, 43 (2011).

[73] M. Eschrig, "Spin-polarized supercurrents for spintronics: a review of current progress", Reports on Progress in Physics **78**, 104501 (2015).

[74] J. Linder and J. W. A. Robinson, "Superconducting spintronics", Nature Physics **11**, 307 (2015).

[75] G. Yang, C. Ciccarelli, and J. W. A. Robinson, "Boosting spintronics with superconductivity", APL Materials **9**, 050703 (2021).

[76] H. B. Heersche, P. Jarillo-Herrero, J. B. Oostinga, L. M. K. Vandersypen, and A. F. Morpurgo, "Bipolar supercurrent in graphene", Nature **446**, 56 (2007).

[77] K. Komatsu, C. Li, S. Autier-Laurent, H. Bouchiat, and S. Guéron, "Superconducting proximity effect in long superconductor/graphene/superconductor junctions: From specular Andreev reflection at zero field to the quantum Hall regime", Physical Review B **86**, 115412 (2012).

[78] V. E. Calado, S. Goswami, G. Nanda, M. Diez, A. R. Akhmerov, K. Watanabe, T. Taniguchi, T. M. Klapwijk, and L. M. K. Vandersypen, "Ballistic Josephson junctions in edge-contacted graphene", Nature Nanotechnology **10**, 761 (2015).

[79] D. I. Indolese, R. Delagrange, P. Makk, J. R. Wallbank, K. Wanatabe, T. Taniguchi, and C. Schönenberger, "Signatures of van Hove Singularities Probed by the Supercurrent in a Graphene-hBN Superlattice", Physical Review Letters **121**, 137701 (2018).

[80] K. Li, X. Feng, W. Zhang, Y. Ou, L. Chen, K. He, L.-L. Wang, L. Guo, G. Liu, Q.-K. Xue, and X. Ma, "Superconductivity in Ca-intercalated epitaxial graphene on silicon carbide", Applied Physics Letters **103**, 062601 (2013).

[81] B. M. Ludbrook, G. Levy, P. Nigge, M. Zonno, M. Schneider, D. J. Dvorak, C. N. Veenstra, S. Zhdanovich, D. Wong, P. Dosanjh, C. Straßer, A. Stöhr, S. Forti, C. R. Ast, U. Starke, and A. Damascelli, "Evidence for superconductivity in Li-decorated monolayer graphene", Proc. Natl. Acad. Sci. U.S.A. **112**, 11795 (2015).

[82] J. Chapman, Y. Su, C. A. Howard, D. Kundys, A. N. Grigorenko, F. Guinea, A. K. Geim, I. V. Grigorieva, and R. R. Nair, "Superconductivity in Ca-doped graphene laminates", Scientific Reports **6**, 23254 (2016).

[83] C. Tonnoir, A. Kimouche, J. Coraux, L. Magaud, B. Delsol, B. Gilles, and C. Chapelier, "Induced Superconductivity in Graphene Grown on Rhenium", Physical Review Letters **111**, 246805 (2013).

[84] A. Di Bernardo, O. Millo, M. Barbone, H. Alpern, Y. Kalcheim, U. Sassi, A. K. Ott, D. De Fazio, D. Yoon, M. Amado, A. C. Ferrari, J. Linder, and J. W. A. Robinson, "p-wave triggered superconductivity in single-layer graphene on an electron-doped oxide superconductor", Nature Communications **8**, 14024 (2017).

[85] J. L. Lado and J. Fernández-Rossier, "Unconventional Yu-Shiba-Rusinov states in hydrogenated graphene", 2D Materials **3**, 025001 (2016).

[86] J. W. McClure, "Band Structure of Graphite and de Haas-van Alphen Effect", Physical Review **108**, 612 (1957).

[87] J. C. Slonczewski and P. R. Weiss, "Band Structure of Graphite", Physical Review **109**, 272 (1958).

[88] S. Konschuh, M. Gmitra, D. Kochan, and J. Fabian, "Theory of spin-orbit coupling in bilayer graphene", Physical Review B **85**, 115423 (2012).

[89] E. McCann and M. Koshino, "The electronic properties of bilayer graphene", Reports on Progress in Physics **76**, 056503 (2013).

[90] D. Kochan, S. Irmer, M. Gmitra, and J. Fabian, "Resonant Scattering by Magnetic Impurities as a Model for Spin Relaxation in Bilayer Graphene", Physical Review Letters **115**, 196601 (2015).

[91] M. Gmitra, D. Kochan, and J. Fabian, "Spin-Orbit Coupling in Hydrogenated Graphene", Physical Review Letters **110**, 246602 (2013).

[92] D. Kochan, M. Gmitra, and J. Fabian, "Spin Relaxation Mechanism in Graphene: Resonant Scattering by Magnetic Impurities", Physical Review Letters **112**, 116602 (2014).

[93] C. R. Harris, K. J. Millman, S. J. van der Walt, R. Gommers, P. Virtanen, D. Cournapeau, E. Wieser, J. Taylor, S. Berg, N. J. Smith, R. Kern, M. Picus, S. Hoyer, M. H. van Kerkwijk, M. Brett, A. Haldane, J. F. del Río, M. Wiebe, P. Peterson, P. Gérard-Marchant, K. Sheppard, T. Reddy, W. Weckesser, H. Abbasi, C. Gohlke, and T. E. Oliphant, "Array programming with NumPy", Nature **585**, 357 (2020).

[94] *NumPy*, https://numpy.org/.

[95] P. Virtanen, R. Gommers, T. E. Oliphant, M. Haberland, T. Reddy, D. Cournapeau, E. Burovski, P. Peterson, W. Weckesser, J. Bright, S. J. van der Walt, M. Brett, J. Wilson, K. J. Millman, N. Mayorov, A. R. J. Nelson, E. Jones, R. Kern, E. Larson, C. J. Carey, İ. Polat, Y. Feng, E. W. Moore, J. VanderPlas, D. Laxalde, J. Perktold, R. Cimrman, I. Henriksen, E. A. Quintero, C. R. Harris, A. M. Archibald, A. H. Ribeiro, F. Pedregosa, P. van Mulbregt, and SciPy 1.0 Contributors, "SciPy 1.0: fundamental algorithms for scientific computing in Python", Nature Methods **17**, 261 (2020); erratum Nature Methods **17**, 352 (2020).

[96] *SciPy*, https://scipy.org/.

[97] J. W. Eaton, D. Bateman, S. Hauberg, and R. Wehbring, *GNU Octave version 6.1.0 manual: a high-level interactive language for numerical computations* (2020).

[98] L. C. Hebel and C. P. Slichter, "Nuclear Relaxation in Superconducting Aluminum", Physical Review **107**, 901 (1957).

[99] L. C. Hebel and C. P. Slichter, "Nuclear Spin Relaxation in Normal and Superconducting Aluminum", Physical Review **113**, 1504 (1959).

[100] L. C. Hebel, "Theory of Nuclear Spin Relaxation in Superconductors", Physical Review **116**, 79 (1959).

[101] J. Fuchs, M. Barth, C. Gorini, İ. Adagideli, and K. Richter, "Crossed Andreev reflection in topological insulator nanowire T junctions", Physical Review B **104**, 085415 (2021).

[102] D. van Delft and P. Kes, "The discovery of superconductivity", Physics Today **63**, 38 (2010); H. Kamerlingh Onnes, Commun. Phys. Lab. Univ. Leiden **120b** (1911); ibid. **122b** (1911); ibid. **124c** (1911).

[103] A. F. Andreev, "Electron Spectrum of the Intermediate State of Superconductors", Zhurnal Eksperimental'noi i Teoreticheskoi Fiziki **49**, 655 (1966); [Soviet Physics JETP **22**, 455 (1966)].

[104] G. Deutscher and D. Feinberg, "Coupling superconducting-ferromagnetic point contacts by Andreev reflections", Applied Physics Letters **76**, 487 (2000).

[105] P. Recher, E. V. Sukhorukov, and D. Loss, "Andreev tunneling, Coulomb blockade, and resonant transport of nonlocal spin-entangled electrons", Physical Review B **63**, 165314 (2001).

[106] J. Schindele, A. Baumgartner, and C. Schönenberger, "Near-Unity Cooper Pair Splitting Efficiency", Physical Review Letters **109**, 157002 (2012).

[107] D. Beckmann, H. B. Weber, and H. v. Löhneysen, "Evidence for Crossed Andreev Reflection in Superconductor-Ferromagnet Hybrid Structures", Physical Review Letters **93**, 197003 (2004).

[108] R. Nehra, D. S. Bhakuni, A. Sharma, and A. Soori, "Enhancement of crossed Andreev reflection in a Kitaev ladder connected to normal metal leads", Journal of Physics: Condensed Matter **31**, 345304 (2019).

[109] H. Haugen, D. Huertas-Hernando, A. Brataas, and X. Waintal, "Crossed Andreev reflection versus electron transfer in three-terminal graphene devices", Physical Review B **81**, 174523 (2010).

[110] R. Beiranvand, H. Hamzehpour, and M. Alidoust, "Nonlocal Andreev entanglements and triplet correlations in graphene with spin-orbit coupling", Physical Review B **96**, 161403 (2017).

[111] M. Beconcini, M. Polini, and F. Taddei, "Nonlocal superconducting correlations in graphene in the quantum Hall regime", Physical Review B **97**, 201403 (2018).

[112] X. Wu, H. Meng, F. Kong, H. Zhang, Y. Bai, and N. Xu, "Tunable nonlocal valley-entangled Cooper pair splitter realized in bilayer-graphene van der Waals spin valves", Physical Review B **101**, 125406 (2020).

[113] S.-B. Zhang and B. Trauzettel, "Perfect Crossed Andreev Reflection in Dirac Hybrid Junctions in the Quantum Hall Regime", Physical Review Letters **122**, 257701 (2019).

[114] M. F. Jakobsen, A. Brataas, and A. Qaiumzadeh, "Electrically Controlled Crossed Andreev Reflection in Two-Dimensional Antiferromagnets", Physical Review Letters **127**, 017701 (2021).

[115]   G. Tkachov and E. M. Hankiewicz, "Spin-helical transport in normal and superconducting topological insulators", Physica Status Solidi B: Basic Research **250**, 215 (2013).

[116]   J. H. Bardarson, P. W. Brouwer, and J. E. Moore, "Aharonov-Bohm Oscillations in Disordered Topological Insulator Nanowires", Physical Review Letters **105**, 156803 (2010).

[117]   Y. Tanaka, M. Sato, and N. Nagaosa, "Symmetry and Topology in Superconductors –Odd-Frequency Pairing and Edge States–", Journal of the Physical Society of Japan **81**, 011013 (2012).

[118]   H. Peng, K. Lai, D. Kong, S. Meister, Y. Chen, X.-L. Qi, S.-C. Zhang, Z.-X. Shen, and Y. Cui, "Aharonov–Bohm interference in topological insulator nanoribbons", Nature Materials **9**, 225 (2010).

[119]   F. Xiu, L. He, Y. Wang, L. Cheng, L.-T. Chang, M. Lang, G. Huang, X. Kou, Y. Zhou, X. Jiang, Z. Chen, J. Zou, A. Shailos, and K. L. Wang, "Manipulating surface states in topological insulator nanoribbons", Nature Nanotechnology **6**, 216 (2011).

[120]   S. S. Hong, J. J. Cha, D. Kong, and Y. Cui, "Ultra-low carrier concentration and surface-dominant transport in antimony-doped $Bi_2Se_3$ topological insulator nanoribbons", Nature Communications **3**, 757 (2012).

[121]   M. Tian, W. Ning, Z. Qu, H. Du, J. Wang, and Y. Zhang, "Dual evidence of surface Dirac states in thin cylindrical topological insulator $Bi_2Te_3$ nanowires", Scientific Reports **3**, 1212 (2013).

[122]   B. Hamdou, J. Gooth, A. Dorn, E. Pippel, and K. Nielsch, "Surface state dominated transport in topological insulator $Bi_2Te_3$ nanowires", Applied Physics Letters **103**, 193107 (2013).

[123]   J. Dufouleur, L. Veyrat, A. Teichgräber, S. Neuhaus, C. Nowka, S. Hampel, J. Cayssol, J. Schumann, B. Eichler, O. G. Schmidt, B. Büchner, and R. Giraud, "Quasiballistic Transport of Dirac Fermions in a $Bi_2Se_3$ Nanowire", Physical Review Letters **110**, 186806 (2013).

[124]   M. Safdar, Q. Wang, M. Mirza, Z. Wang, K. Xu, and J. He, "Topological Surface Transport Properties of Single-Crystalline SnTe Nanowire", Nano Letters **13**, 5344 (2013).

[125]  Z. Wang, R. L. J. Qiu, C. H. Lee, Z. Zhang, and X. P. A. Gao, "Ambipolar Surface Conduction in Ternary Topological Insulator $Bi_2(Te_{1-x}Se_x)_3$ Nanoribbons", ACS Nano **7**, 2126 (2013).

[126]  S. S. Hong, Y. Zhang, J. J. Cha, X.-L. Qi, and Y. Cui, "One-Dimensional Helical Transport in Topological Insulator Nanowire Interferometers", Nano Letters **14**, 2815 (2014).

[127]  S. Bäßler, B. Hamdou, P. Sergelius, A.-K. Michel, R. Zierold, H. Reith, J. Gooth, and K. Nielsch, "One-dimensional edge transport on the surface of cylindrical $Bi_xTe_{3-y}Se_y$ nanowires in transverse magnetic fields", Applied Physics Letters **107**, 181602 (2015).

[128]  S. Cho, B. Dellabetta, R. Zhong, J. Schneeloch, T. Liu, G. Gu, M. J. Gilbert, and N. Mason, "Aharonov–Bohm oscillations in a quasi-ballistic three-dimensional topological insulator nanowire", Nature Communications **6**, 7634 (2015).

[129]  Y. C. Arango, L. Huang, C. Chen, J. Avila, M. C. Asensio, D. Grützmacher, H. Lüth, J. G. Lu, and T. Schäpers, "Quantum Transport and Nano Angle-resolved Photoemission Spectroscopy on the Topological Surface States of Single $Sb_2Te_3$ Nanowires", Scientific Reports **6**, 29493 (2016).

[130]  L. A. Jauregui, M. T. Pettes, L. P. Rokhinson, L. Shi, and Y. P. Chen, "Magnetic field-induced helical mode and topological transitions in a topological insulator nanoribbon", Nature Nanotechnology **11**, 345 (2016).

[131]  B. Bhattacharyya, A. Sharma, V. P. S. Awana, T. D. Senguttuvan, and S. Husale, "FIB synthesis of $Bi_2Se_3$ 1D nanowires demonstrating the co-existence of Shubnikov–de Haas oscillations and linear magnetoresistance", Journal of Physics: Condensed Matter **29**, 07LT01 (2016).

[132]  J. Dufouleur, L. Veyrat, B. Dassonneville, E. Xypakis, J. H. Bardarson, C. Nowka, S. Hampel, J. Schumann, B. Eichler, O. G. Schmidt, B. Büchner, and R. Giraud, "Weakly-coupled quasi-1D helical modes in disordered 3D topological insulator quantum wires", Scientific Reports **7**, 45276 (2017).

[133]  J. Ziegler, R. Kozlovsky, C. Gorini, M.-H. Liu, S. Weishäupl, H. Maier, R. Fischer, D. A. Kozlov, Z. D. Kvon, N. Mikhailov, S. A. Dvoretsky, K. Richter, and D. Weiss, "Probing spin helical surface states in topological HgTe nanowires", Physical Review B **97**, 035157 (2018).

[134] G. Kunakova, L. Galletti, S. Charpentier, J. Andzane, D. Erts, F. Léonard, C. D. Spataru, T. Bauch, and F. Lombardi, "Bulk-free topological insulator $Bi_2Se_3$ nanoribbons with magnetotransport signatures of Dirac surface states", Nanoscale **10**, 19595 (2018).

[135] D. Rosenbach, N. Oellers, A. R. Jalil, M. Mikulics, J. Kölzer, E. Zimmermann, G. Mussler, S. Bunte, D. Grützmacher, H. Lüth, and T. Schäpers, "Quantum Transport in Topological Surface States of Selectively Grown $Bi_2Te_3$ Nanoribbons", Advanced Electronic Materials **6**, 2000205 (2020).

[136] F. Münning, O. Breunig, H. F. Legg, S. Roitsch, D. Fan, M. Rößler, A. Rosch, and Y. Ando, "Quantum confinement of the Dirac surface states in topological-insulator nanowires", Nat. Commun. **12**, 1038 (2021).

[137] D. Rosenbach, K. Moors, A. R. Jalil, J. Kölzer, E. Zimmermann, J. Schubert, S. Karimzadah, G. Mussler, P. Schüffelgen, D. Grützmacher, H. Lüth, and T. Schäpers, "Gate-induced decoupling of surface and bulk state properties in selectively-deposited $Bi_2Te_3$ nanoribbons", SciPost Physics Core **5**, 17 (2022).

[138] H. F. Legg, M. Rößler, F. Münning, D. Fan, O. Breunig, A. Bliesener, G. Lippertz, A. Uday, A. A. Taskin, D. Loss, J. Klinovaja, and Y. Ando, "Giant magnetochiral anisotropy from quantum-confined surface states of topological insulator nanowires", Nature Nanotechnology, 10.1038/s41565-022-01124-1 (2022).

[139] J. Nilsson, A. R. Akhmerov, and C. W. J. Beenakker, "Splitting of a Cooper Pair by a Pair of Majorana Bound States", Physical Review Letters **101**, 120403 (2008).

[140] A. R. Akhmerov, J. Nilsson, and C. W. J. Beenakker, "Electrically Detected Interferometry of Majorana Fermions in a Topological Insulator", Physical Review Letters **102**, 216404 (2009).

[141] L. Fu and C. L. Kane, "Probing Neutral Majorana Fermion Edge Modes with Charge Transport", Physical Review Letters **102**, 216403 (2009).

[142] K. Zhang and Q. Cheng, "Electrically tunable crossed Andreev reflection in a ferromagnet–superconductor–ferromagnet junction on a topological insulator", Superconductor Science and Technology **31**, 075001 (2018).

[143] F. Keidel, S.-Y. Hwang, B. Trauzettel, B. Sothmann, and P. Burset, "On-demand thermoelectric generation of equal-spin Cooper pairs", Physical Review Research **2**, 022019 (2020).

[144] G. Blasi, F. Taddei, V. Giovannetti, and A. Braggio, "Manipulation of Cooper pair entanglement in hybrid topological Josephson junctions", Physical Review B **99**, 064514 (2019).

[145] W. Chen, R. Shen, L. Sheng, B. G. Wang, and D. Y. Xing, "Resonant nonlocal Andreev reflection in a narrow quantum spin Hall system", Physical Review B **84**, 115420 (2011).

[146] R. W. Reinthaler, P. Recher, and E. M. Hankiewicz, "Proposal for an All-Electrical Detection of Crossed Andreev Reflection in Topological Insulators", Physical Review Letters **110**, 226802 (2013).

[147] F. Crépin, P. Burset, and B. Trauzettel, "Odd-frequency triplet superconductivity at the helical edge of a topological insulator", Physical Review B **92**, 100507 (2015).

[148] C. Fleckenstein, N. T. Ziani, and B. Trauzettel, "Conductance signatures of odd-frequency superconductivity in quantum spin Hall systems using a quantum point contact", Physical Review B **97**, 134523 (2018).

[149] J. Cayssol, "Crossed Andreev Reflection in a Graphene Bipolar Transistor", Physical Review Letters **100**, 147001 (2008).

[150] D. Breunig, P. Burset, and B. Trauzettel, "Creation of Spin-Triplet Cooper Pairs in the Absence of Magnetic Ordering", Physical Review Letters **120**, 037701 (2018).

[151] S. F. Islam, P. Dutta, and A. Saha, "Enhancement of crossed Andreev reflection in a normal-superconductor-normal junction made of thin topological insulator", Physical Review B **96**, 155429 (2017).

[152] F. de Juan, R. Ilan, and J. H. Bardarson, "Robust Transport Signatures of Topological Superconductivity in Topological Insulator Nanowires", Physical Review Letters **113**, 107003 (2014).

[153] J. H. Bardarson and J. E. Moore, "Quantum interference and Aharonov–Bohm oscillations in topological insulators", Reports on Progress in Physics **76**, 056501 (2013).

[154] J. H. Bardarson and R. Ilan, "Transport in Topological Insulator Nanowires", in *Topological Matter: Lectures from the Topological Matter School 2017*, edited by D. Bercioux, J. Cayssol, M. G. Vergniory, and M. Reyes Calvo (Springer International Publishing, Cham, 2018), pp. 93–114.

[155] Y. Zhang, Y. Ran, and A. Vishwanath, "Topological insulators in three dimensions from spontaneous symmetry breaking", Physical Review B **79**, 245331 (2009).

[156] Y. Zhang and A. Vishwanath, "Anomalous Aharonov-Bohm Conductance Oscillations from Topological Insulator Surface States", Physical Review Letters **105**, 206601 (2010).

[157] R. Ilan, J. H. Bardarson, H.-S. Sim, and J. E. Moore, "Detecting perfect transmission in Josephson junctions on the surface of three dimensional topological insulators", New Journal of Physics **16**, 053007 (2014).

[158] E. Xypakis and J. H. Bardarson, "Conductance fluctuations and disorder induced $v = 0$ quantum Hall plateau in topological insulator nanowires", Physical Review B **95**, 035415 (2017).

[159] E. Xypakis, J.-W. Rhim, J. H. Bardarson, and R. Ilan, "Perfect transmission and Aharanov-Bohm oscillations in topological insulator nanowires with nonuniform cross section", Physical Review B **101**, 045401 (2020).

[160] K. Moors, P. Schüffelgen, D. Rosenbach, T. Schmitt, T. Schäpers, and T. L. Schmidt, "Magnetotransport signatures of three-dimensional topological insulator nanostructures", Physical Review B **97**, 245429 (2018); erratum Physical Review B **103**, 079902 (2021).

[161] R. Kozlovsky, "Magnetotransport in 3D Topological Insulator Nanowires", Dissertation (Universität Regensburg, Apr. 2020).

[162] Y. Ran, A. Vishwanath, and D.-H. Lee, "Spin-Charge Separated Solitons in a Topological Band Insulator", Physical Review Letters **101**, 086801 (2008).

[163] D.-H. Lee, "Surface States of Topological Insulators: The Dirac Fermion in Curved Two-Dimensional Spaces", Physical Review Letters **103**, 196804 (2009).

[164] F. de Juan, J. H. Bardarson, and R. Ilan, "Conditions for fully gapped topological superconductivity in topological insulator nanowires", SciPost Physics **6**, 60 (2019).

[165] L. Brey and H. A. Fertig, "Electronic states of wires and slabs of topological insulators: Quantum Hall effects and edge transport", Physical Review B **89**, 085305 (2014).

[166] Y.-Y. Zhang, X.-R. Wang, and X. C. Xie, "Three-dimensional topological insulator in a magnetic field: chiral side surface states and quantized Hall conductance", Journal of Physics: Condensed Matter **24**, 015004 (2011).

[167] E. J. König, P. M. Ostrovsky, I. V. Protopopov, I. V. Gornyi, I. S. Burmistrov, and A. D. Mirlin, "Half-integer quantum Hall effect of disordered Dirac fermions at a topological insulator surface", Physical Review B **90**, 165435 (2014).

[168] C. W. Groth, M. Wimmer, A. R. Akhmerov, and X. Waintal, "Kwant: a software package for quantum transport", New Journal of Physics **16**, 063065 (2014).

[169] M. Wimmer, A. R. Akhmerov, J. P. Dahlhaus, and C. W. J. Beenakker, "Quantum point contact as a probe of a topological superconductor", New Journal of Physics **13**, 053016 (2011).

[170] B. Béri, "Dephasing-enabled triplet Andreev conductance", Physical Review B **79**, 245315 (2009).

[171] *Kwant*, https://kwant-project.org/.

[172] P. R. Amestoy, I. S. Duff, J.-Y. L'Excellent, and J. Koster, "A Fully Asynchronous Multifrontal Solver Using Distributed Dynamic Scheduling", SIAM Journal on Matrix Analysis and Applications **23**, 15 (2001).

[173] P. R. Amestoy, A. Buttari, J.-Y. L'Excellent, and T. Mary, "Performance and Scalability of the Block Low-Rank Multifrontal Factorization on Multicore Architectures", ACM Transactions on Mathematical Software **45**, 1 (2019).

[174] H. Zhang, C.-X. Liu, X.-L. Qi, X. Dai, Z. Fang, and S.-C. Zhang, "Topological insulators in $Bi_2Se_3$, $Bi_2Te_3$ and $Sb_2Te_3$ with a single Dirac cone on the surface", Nature Physics **5**, 438 (2009).

[175] C. J. Lambert and R. Raimondi, "Phase-coherent transport in hybrid superconducting nanostructures", Journal of Physics: Condensed Matter **10**, 901 (1998).

[176] C.-X. Liu, X.-L. Qi, H. Zhang, X. Dai, Z. Fang, and S.-C. Zhang, "Model Hamiltonian for topological insulators", Physical Review B **82**, 045122 (2010).

[177] J. Wiedenmann, E. Bocquillon, R. S. Deacon, S. Hartinger, O. Herrmann, T. M. Klapwijk, L. Maier, C. Ames, C. Brüne, C. Gould, A. Oiwa, K. Ishibashi, S. Tarucha, H. Buhmann, and L. W. Molenkamp, "$4\pi$-periodic Josephson supercurrent in HgTe-based topological Josephson junctions", Nature Communications **7**, 10303 (2016).

[178] H. J. Suominen, J. Danon, M. Kjaergaard, K. Flensberg, J. Shabani, C. J. Palmstrøm, F. Nichele, and C. M. Marcus, "Anomalous Fraunhofer interference in epitaxial superconductor-semiconductor Josephson junctions", Physical Review B **95**, 035307 (2017).

[179] K. Gharavi, G. W. Holloway, R. R. LaPierre, and J. Baugh, "Nb/InAs nanowire proximity junctions from Josephson to quantum dot regimes", Nanotechnology **28**, 085202 (2017).

[180] K. Zuo, V. Mourik, D. B. Szombati, B. Nijholt, D. J. van Woerkom, A. Geresdi, J. Chen, V. P. Ostroukh, A. R. Akhmerov, S. R. Plissard, D. Car, E. P. A. M. Bakkers, D. I. Pikulin, L. P. Kouwenhoven, and S. M. Frolov, "Supercurrent Interference in Few-Mode Nanowire Josephson Junctions", Physical Review Letters **119**, 187704 (2017).

[181] A. Q. Chen, M. J. Park, S. T. Gill, Y. Xiao, D. Reig-i-Plessis, G. J. MacDougall, M. J. Gilbert, and N. Mason, "Finite momentum Cooper pairing in three-dimensional topological insulator Josephson junctions", Nature Communications **9**, 3478 (2018).

[182] C.-Z. Li, A.-Q. Wang, C. Li, W.-Z. Zheng, A. Brinkman, D.-P. Yu, and Z.-M. Liao, "Topological Transition of Superconductivity in Dirac Semimetal Nanowire Josephson Junctions", Physical Review Letters **126**, 027001 (2021).

[183] A. Kringhøj, G. W. Winkler, T. W. Larsen, D. Sabonis, O. Erlandsson, P. Krogstrup, B. van Heck, K. D. Petersson, and C. M. Marcus, "Andreev Modes from Phase Winding in a Full-Shell Nanowire-Based Transmon", Physical Review Letters **126**, 047701 (2021).

[184] P. Perla, H. A. Fonseka, P. Zellekens, R. Deacon, Y. Han, J. Kölzer, T. Mörstedt, B. Bennemann, A. Espiari, K. Ishibashi, D. Grützmacher, A. M. Sanchez, M. I. Lepsa, and T. Schäpers, "Fully in situ Nb/InAs-nanowire Josephson junctions by selective-area growth and shadow evaporation", Nanoscale Advances **3**, 1413 (2021).

[185] L. Stampfer, D. J. Carrad, D. Olsteins, C. E. N. Petersen, S. A. Khan, P. Krogstrup, and T. S. Jespersen, "Andreev Interference in the Surface Accumulation Layer of Half-Shell InAsSb/Al Hybrid Nanowires", Advanced Materials **34**, 2108878 (2022).

[186] H. F. Legg, D. Loss, and J. Klinovaja, "Metallization and proximity superconductivity in topological insulator nanowires", Physical Review B **105**, 155413 (2022).

[187] V. P. Ostroukh, B. Baxevanis, A. R. Akhmerov, and C. W. J. Beenakker, "Two-dimensional Josephson vortex lattice and anomalously slow decay of the Fraunhofer oscillations in a ballistic SNS junction with a warped Fermi surface", Physical Review B **94**, 094514 (2016).

[188] I. O. Kulik and A. N. Omel'Yanchuk, "Properties of superconducting microbridges in the pure limit", Fizika Nizkikh Temperatur **3**, 945 (1977); [Sov. J. Low Temp. Phys. **3**, 459 (1977)].

[189] C. W. J. Beenakker and H. van Houten, "Josephson current through a superconducting quantum point contact shorter than the coherence length", Physical Review Letters **66**, 3056 (1991).

[190] F. Pientka, A. Keselman, E. Berg, A. Yacoby, A. Stern, and B. I. Halperin, "Topological Superconductivity in a Planar Josephson Junction", Physical Review X **7**, 021032 (2017).

[191] C. W. J. Beenakker, "Three "Universal" Mesoscopic Josephson Effects", in Transport Phenomena in Mesoscopic Systems, edited by H. Fukuyama and T. Ando (1992), pp. 235–253.

[192] C. Beenakker and H. van Houten, "THE SUPERCONDUCTING QUANTUM POINT CONTACT", in *Nanostructures and Mesoscopic Systems*, edited by W. P. Kirk and M. A. Reed (Academic Press, 1992), pp. 481–497.

[193] T. Yokoyama, M. Eto, and Y. V. Nazarov, "Josephson Current through Semiconductor Nanowire with Spin–Orbit Interaction in Magnetic Field", Journal of the Physical Society of Japan **82**, 054703 (2013).

[194] T. Yokoyama, M. Eto, and Y. V. Nazarov, "Anomalous Josephson effect induced by spin-orbit interaction and Zeeman effect in semiconductor nanowires", Physical Review B **89**, 195407 (2014).

[195] T. Yokoyama, M. Eto, and Y. V. Nazarov, "Critical current in semiconductor nanowire Josephson junctions in the presence of magnetic field", Journal of Physics: Conference Series **568**, 052035 (2014).

[196] P. Sriram, S. S. Kalantre, K. Gharavi, J. Baugh, and B. Muralidharan, "Supercurrent interference in semiconductor nanowire Josephson junctions", Physical Review B **100**, 155431 (2019).

[197] G. E. Blonder, M. Tinkham, and T. M. Klapwijk, "Transition from metallic to tunneling regimes in superconducting microconstrictions: Excess current, charge imbalance, and supercurrent conversion", Physical Review B **25**, 4515 (1982).

[198] R. Fischer, W. Himmler, D. Weiss, D. A. Kozlov, N. N. Mikhailov, S. A. Dvoretsky, M. Barth, J. Fuchs, C. Gorini, and K. Richter, "Supercurrent oscillations in HgTe Josephson junctions", unpublished, 2022.

[199] Y. Cao, V. Fatemi, A. Demir, S. Fang, S. L. Tomarken, J. Y. Luo, J. D. Sanchez-Yamagishi, K. Watanabe, T. Taniguchi, E. Kaxiras, R. C. Ashoori, and P. Jarillo-Herrero, "Correlated insulator behaviour at half-filling in magic-angle graphene superlattices", Nature **556**, 80 (2018).

[200] Y. Cao, V. Fatemi, S. Fang, K. Watanabe, T. Taniguchi, E. Kaxiras, and P. Jarillo-Herrero, "Unconventional superconductivity in magic-angle graphene superlattices", Nature **556**, 43 (2018).

[201] E. Y. Andrei and A. H. MacDonald, "Graphene bilayers with a twist", Nature Materials **19**, 1265 (2020).

[202] Z. Hao, A. M. Zimmerman, P. Ledwith, E. Khalaf, D. H. Najafabadi, K. Watanabe, T. Taniguchi, A. Vishwanath, and P. Kim, "Electric field-tunable superconductivity in alternating-twist magic-angle trilayer graphene", Science **371**, 1133 (2021).

[203] J. M. Park, Y. Cao, K. Watanabe, T. Taniguchi, and P. Jarillo-Herrero, "Tunable strongly coupled superconductivity in magic-angle twisted trilayer graphene", Nature **590**, 249 (2021).

[204] H. Zhou, T. Xie, T. Taniguchi, K. Watanabe, and A. F. Young, "Superconductivity in rhombohedral trilayer graphene", Nature **598**, 434 (2021).

[205] J. S. Bell, "On the Einstein Podolsky Rosen paradox", Physics Physique Fizika **1**, 195 (1964).

[206] F. Wilczek, "Majorana returns", Nature Phys. **5**, 614 (2009).

[207] V. Mourik, K. Zuo, S. M. Frolov, S. R. Plissard, E. P. A. M. Bakkers, and L. P. Kouwenhoven, "Signatures of Majorana Fermions in Hybrid Superconductor-Semiconductor Nanowire Devices", Science **336**, 1003 (2012).

[208] L. P. Rokhinson, X. Liu, and J. K. Furdyna, "The fractional a.c. Josephson effect in a semiconductor–superconductor nanowire as a signature of Majorana particles", Nature Physics **8**, 795 (2012).

[209]  M. T. Deng, S. Vaitiekėnas, E. B. Hansen, J. Danon, M. Leijnse, K. Flensberg, J. Nygård, P. Krogstrup, and C. M. Marcus, "Majorana bound state in a coupled quantum-dot hybrid-nanowire system", Science **354**, 1557 (2016).

[210]  S. M. Albrecht, A. P. Higginbotham, M. Madsen, F. Kuemmeth, T. S. Jespersen, J. Nygård, P. Krogstrup, and C. M. Marcus, "Exponential protection of zero modes in Majorana islands", Nature **531**, 206 (2016).

[211]  Ö. Gül, Y. Ronen, S. Y. Lee, H. Shapourian, J. Zauberman, Y. H. Lee, K. Watanabe, T. Taniguchi, A. Vishwanath, A. Yacoby, and P. Kim, "Induced superconductivity in the fractional quantum Hall edge", 10.48550/arXiv.2009.07836 (2020).

[212]  D. Castelvecchi, "Evidence of elusive Majorana particle dies — but computing hope lives on", Nature **591**, 354 (2021).

[213]  H. Zhang, M. W. A. de Moor, J. D. S. Bommer, D. Xu, G. Wang, N. van Loo, C.-X. Liu, S. Gazibegovic, J. A. Logan, D. Car, R. L. M. O. h. Veld, P. J. van Veldhoven, S. Koelling, M. A. Verheijen, M. Pendharkar, D. J. Pennachio, B. Shojaei, J. S. Lee, C. J. Palmstrøm, E. P. A. M. Bakkers, S. D. Sarma, and L. P. Kouwenhoven, "Large zero-bias peaks in InSb-Al hybrid semiconductor-superconductor nanowire devices", 10.48550/arXiv.2101.11456 (2021).

[214]  H. Song, Z. Zhang, D. Pan, D. Liu, Z. Wang, Z. Cao, L. Liu, L. Wen, D. Liao, R. Zhuo, D. E. Liu, R. Shang, J. Zhao, and H. Zhang, "Large zero bias peaks and dips in a four-terminal thin InAs-Al nanowire device", 10.48550/arXiv.2107.08282 (2021).

[215]  H. F. Legg, D. Loss, and J. Klinovaja, "Majorana bound states in topological insulators without a vortex", Physical Review B **104**, 165405 (2021).

[216]  H. Nielsen and M. Ninomiya, "A no-go theorem for regularizing chiral fermions", Physics Letters B **105**, 219 (1981).

[217]  H. Nielsen and M. Ninomiya, "Absence of neutrinos on a lattice: (I). Proof by homotopy theory", Nuclear Physics B **185**, 20 (1981).

[218]  H. Nielsen and M. Ninomiya, "Absence of neutrinos on a lattice: (II). Intuitive topological proof", Nuclear Physics B **193**, 173 (1981).

[219]  K. M. M. Habib, R. N. Sajjad, and A. W. Ghosh, "Modified Dirac Hamiltonian for efficient quantum mechanical simulations of micron sized devices", Applied Physics Letters **108**, 113105 (2016).

[220]  K. G. Wilson, "Confinement of quarks", Physical Review D **10**, 2445 (1974).

[221]  L. Susskind, "Lattice fermions", Physical Review D **16**, 3031 (1977).

[222]  R. Stacey, "Eliminating lattice fermion doubling", Physical Review D **26**, 468 (1982).

[223]  A. R. Hernández and C. H. Lewenkopf, "Finite-difference method for transport of two-dimensional massless Dirac fermions in a ribbon geometry", Physical Review B **86**, 155439 (2012).

[224]  S. Hong, V. Diep, S. Datta, and Y. P. Chen, "Modeling potentiometric measurements in topological insulators including parallel channels", Physical Review B **86**, 085131 (2012).

[225]  M. J. Pacholski, G. Lemut, J. Tworzydło, and C. W. J. Beenakker, "Generalized eigenproblem without fermion doubling for Dirac fermions on a lattice", SciPost Physics **11**, 105 (2021).

[226]  R. Peierls, "Zur Theorie des Diamagnetismus von Leitungselektronen", Zeitschrift für Physik **80**, 763 (1933).

[227]  P. Wójcik and M. P. Nowak, "Durability of the superconducting gap in Majorana nanowires under orbital effects of a magnetic field", Physical Review B **97**, 235445 (2018).

[228]  G. W. Winkler, A. E. Antipov, B. van Heck, A. A. Soluyanov, L. I. Glazman, M. Wimmer, and R. M. Lutchyn, "Unified numerical approach to topological semiconductor-superconductor heterostructures", Physical Review B **99**, 245408 (2019).

[229]  J. Fuchs, J. Main, H. Cartarius, and G. Wunner, "Harmonic inversion analysis of exceptional points in resonance spectra", Journal of Physics A: Mathematical and Theoretical **47**, 125304 (2014).

[230]  J. Brinker, J. Fuchs, J. Main, G. Wunner, and H. Cartarius, "Verification of exceptional points in the collapse dynamics of Bose-Einstein condensates", Physical Review A **91**, 013609 (2015).

# List of publications

[24]  M. Barth, J. Fuchs, and D. Kochan, "Spin relaxation, Josephson effect, and Yu-Shiba-Rusinov states in superconducting bilayer graphene", Physical Review B **105**, 205409 (2022).

[101]  J. Fuchs, M. Barth, C. Gorini, İ. Adagideli, and K. Richter, "Crossed Andreev reflection in topological insulator nanowire T junctions", Physical Review B **104**, 085415 (2021).

[198]  R. Fischer, W. Himmler, D. Weiss, D. A. Kozlov, N. N. Mikhailov, S. A. Dvoretsky, M. Barth, J. Fuchs, C. Gorini, and K. Richter, "Supercurrent oscillations in HgTe Josephson junctions", unpublished, 2022.

[229]  J. Fuchs, J. Main, H. Cartarius, and G. Wunner, "Harmonic inversion analysis of exceptional points in resonance spectra", Journal of Physics A: Mathematical and Theoretical **47**, 125304 (2014).

[230]  J. Brinker, J. Fuchs, J. Main, G. Wunner, and H. Cartarius, "Verification of exceptional points in the collapse dynamics of Bose-Einstein condensates", Physical Review A **91**, 013609 (2015).

# List of Figures

# List of Acronyms

| | |
|-----|-----|
| ABS | Andreev bound state |
| AR | Andreev reflection |
| BLG | bilayer graphene |
| CAR | crossed Andreev reflection |
| DOS | density of states |
| LDOS | local density of states |
| R | reflection |
| S | superconductor |
| SBLG | superconducting bilayer graphene |
| T | transmission |
| TI | topological insulator |
| YSR | Yu-Shiba-Rusinov |
| 1D | one-dimensional |
| 2D | two-dimensional |
| 3D | three-dimensional |

# Acknowledgments

First and foremost I thank my supervisor Klaus Richter for giving me the opportunity to write this thesis and work in his research group. He introduced me to the fascinating physics of topological insulators and allowed me to dive into the realms of mesoscopic and solid state physics. I am thankful for the opportunities to visit graduate schools and conferences. Last but not least, I appreciate very much that he showed consideration for me being ill for a longer time and taking care of my health.

I want to thank all friends and family members who visited me in the hospital and took care of me when recovering from my disease, among them Andreas A., Sarah, Peter, Johanna, Paul, Friederike, Andreas L., Jörg, my parents Peter and Cornelia, and my sister Henrike, to mention a few of them explicitly. Without them, I would not have returned so well for writing my thesis (if at all).

Next, I thank everyone who proofread this thesis, namely Klaus, Denis, Michael, Vanessa, Gesa, and Andreas C.

Over the last years, the research group has become a quite familiar place. I am very grateful to all colleagues who provided a nice welcome upon arriving in Regensburg and to all group members for the nice atmosphere in general.

I am happy about all enriching physics discussions I had during my research, namely the ones with Klaus, Cosimo, Michael, Raphael, Denis, Andreas C., İnanç, Ralf, and Wolfgang, as well as for all other people I could learn from.

Special thanks goes to Michael for the close and fruitful collaboration. In nearly all projects, I enjoyed working together with him and his more practical yet very physical and sensible attitude. I also appreciate the collaboration with the experimental colleagues from Regensburg, Dieter Weiss, Ralf, and Wolfgang.

I thank all other system administrators, especially Josef, Raphael, Michael, Vanessa and Fabian: Working with them and taking care of the computers of the institute together was a lot of fun. I want to mention Josef in particular who familiarized me with this work.

Special thanks goes to the all the secretaries who worked at our group, namely Toni, Doris, and Elke: They have been a big help in all the depths of the university bureaucracy.

Finally, I acknowledge funding from the Deutsche Forschungsgemeinschaft (DFG, German Research Foundation) within SFB1277 (Project No. A07) and from the Elite Network of Bavaria within the graduate school "Topological Insulators".